ADVENTURE IN THE AIR
Memoirs of a Flight Test Engineer

Maurice Patrick Rose'Meyer

Copyright © 2009 by Maurice Patrick Rose 'Meyer

All rights Reserved

No part of this book may be reproduced, stored in a retrieval system, or transmitted by any means, electronic, mechanical, photocopying, recording, or otherwise, without written permission from the author or publisher. There is one exception. Brief passages may be quoted in articles or reviews.

Library and Archives Canada Cataloguing in Publication

CIP data on file with the National Library and Archives

ISBN 978-1-926582-14-6

FOREWORD
By Wally Warner

I must be honest and state that this is my first "Foreward" to any literary work. It is fair to say that this is also Maurice Rose"Meyer's or Rosie's to those of us who have known and respected him for many years, first book and in my opinion, he has set the bar high. It is therefore my challenge to write a worthy "Forward" to this most interesting life story.

I first met Maurice in the mid seventies during my tenure as Chief Pilot for the Ontario Ministry of Natural Resources and began what was to become a career altering association with de Havilland and a lifelong friendship with Maurice. We had been using the de Havilland Twin Otter for water bombing fires for several years and their effectiveness was limited by the fuel versus a water load that could be carried with the certified gross take off weight of 12,500 pounds. The possibilities of increasing the maximum take off weight was discussed with de Havilland and a plan was formulated which included a flight test program, to produce data to be presented to the Department of Transport, (known today as Transport Canada) for approval of the water bombing weight increase. As recounted during Maurice's de Havilland years, I became involved in the performance part of the program, due to de Havilland's Chief Test Pilot, Bob Fowler being unavailable. Maurice, in his position as Chief Flight Test Engineer, was instrumental in setting fourth the test program to be flown and of course was on the site to oversee the test activity. As is portrayed throughout the book, Maurice approaches each and every task with great enthusiasm. I was impressed with his attention to detail on this test program. I was a true novice with respect to engineering test flying and I greatly appreciated his support and direction.

The expanding of the Weight and water payload of the increased weight Twin Otter, at Sault Ste Marie, illustrates Rosie's dedication to flight testing, as he accompanied me in the co-pilot's seat for the early attempts at downwind, into current take offs. Maurice showed no signs of concern over our close proximity to the rapids of the Saint Mary's

River at the conclusion of our take off aborts. A lesser man would have asked to be excused from participating in any additional take off attempts.

But as Maurice recounts, he "stayed the course" and we successfully completed the test program.

As de Havilland's Chief Engineering Test Pilot, I had the pleasure of working closely with Maurice on many programs including the ice reconnaissance Dash 7, Coastal Patrol Twin Otter and the Dash 8 Series 100/200/300 aircraft. What is very clear in our business is that safety must, above all other considerations, be paramount. This is achieved through risk management, which is based on flight test knowledge and experience by those responsible for flight test programs. I was blessed to have had the benefit of Rosie's years of experience and my flight test achievements, both during my working directly with him and in the years after his retirement, are a tribute to his ability and wisdom. As the reader is first exposed to a young Master Rose'Meyer growing up in the pre World War Two era in India, including his formal grade school education, early teen years, then on to college and into the working world, Maurice describes all aspects of his life with humor and humility.

Maurice has been a mentor and a life long friend. He represents what is best in aviation, "dedication, enthusiasm and accomplishment".

It has been an honor to review his first public literary effort and I highly recommend his book as a source of enjoyment and education.

This book is dedicated to eight of my associates who lost their lives conducting test flights on both civil and military aircraft and in memory of the large numbers of the Flight Test Community that I did not know, who gave their lives to make aircraft safer. I especially remember the six members of the flight test department of the Handley Page Aircraft Company, who were killed in a Victor V bomber, conducting airspeed error tests at a low altitude.

I salute the pioneer aviators, like George Neal, who in 1960, bailed out of a de Havilland Caribou when the control wheel came loose in his hands, as a result of elevator control flutter. In his ninety first year, George still flyes his beloved DH Chipmunk aircraft.

I salute also those in the Flight Test community, who have either bailed out by parachute, some more than once, and faced life-threatening experiences but have pressed on. They have continued their careers, no doubt, because their chosen profession provided adventure, travel and a thirst to solve problems. To all those, I also dedicate this book.

AUTHORS' NOTES

A famed quote from Lord Brabazon of Tara, a pioneer aviator, is as follows: - "I take the view, and always have that if you cannot say what you are going to say in twenty minutes you ought to go away and write a book." Hence ! This book.

This book is also written in response to many understandable blank stares directed to many colleagues, and me when a technical problem encountered during test flying was discussed with family and friends.

In order to bring home to others the exciting world of test flying, in a non-technical manner, I have chosen to introduce the subject through life experiences and personnel encounters with threatening situations.

With this approach, I hope to lure the reader into the world of aviation during the hay days of aircraft research and development, that followed the Second World War and was fueled by the urgency created by the nuclear threat during the cold war. I wish also to show the effort that goes into making aircraft safe and leave the reader in awe of magnificent men and their flying machines.

When traveling in the safety and comfort of current transport aircraft, I hope the reader will gain some insight into the effort that goes into making aircraft safe. I sincerely hope that by gaining some understanding of the extent of flight testing undertaken to obtain a certificate of Airworthiness, the fear of flying can be alleviated, in those that need assurance.

Some dates and numbers quoted, may not be exact, but close enough, as not to matter.

CHAPTER 1
EARLY DAYS (1928-1947)

My interest in aircraft, started on an RAF dirt airstrip in Peshawar, then in British India, now in Pakistan. It was in the year 1939. I was 11 years old at the time and stood for hours in the hot sun, soaking up the sights and sounds of Hawker Hart biplanes. The fighter planes were about 50 feet from the crude barbed wire fence where I paced up and down trying to get a better view. I could see the machine guns mounted in front of the pilots. They took off and landed with lots of dust, sometimes marring my view. Occasionally a pilot would wave to me from his cockpit, what a thrill it was for one so young. Little did I know that, ten decades later, I would be at a dirt strip in Arizona, directing tests that would enable a Dash 8 aircraft to operate in desert strips in Saudi Arabia.

The word, cockpit, always intrigued me. It stems from a confined space, where two fighting gamecocks were placed in order to fight and entertain a crowd of bloodthirsty onlookers. In this arena they would attack each other, sometimes to death. The cockpit in a fighter aircraft is also a confined space from which a pilot does battle with his enemy.

When aircraft, battle in the air, it is termed a dogfight, not a cockfight.

So why is he not in a dog-pit?

Those planes I saw taking off, no doubt, were off to attack one of the many tribal encampments in Afghanistan in an effort to protect the Indian border.

My uncle, who was a Doctor and Surgeon in the Indian Army, was based in Peshawar, located in the North West Frontier Province. There were, and still are, conflicts among the many tribes in that area of that Province, which bordered Afghanistan. Conflict appeared to be the way of life in that lawless region and most tribesmen were well armed with an inbred instinct to be rebellious.

We were staying adjacent to an army base, where I was told that the soldiers would go to sleep with their rifles chained to their ankles. This was because the Afghans would come down from the hills at night, to steal rifles from the armory. Chains would not always stop the theft.

Men sleeping in their barracks were found with their legs chopped off and the rifles gone.

One day in Peshawar, the ground was alive with locusts and everything was covered with them. Even the streets appeared to be mobile, with hoards of these large grasshoppers crawling about. Little did I know then that years later I would be involved with developing and testing baffles and screens on a Beaver aircraft to protect the engine and oil cooler, during flights into clouds of locusts and spraying them with pesticides. Perhaps my small contribution may have led to the saving of some crops and reduced starvation somewhere in Asia, Africa, Australia and the many other places still ravaged by locust plagues.

A trip by train from Peshawar through the Khyber Pass was one of the highlights of this visit. The journey commenced from Peshawar and ended in Landicotal in Afghanistan. The rugged terrain through the pass gave some indication why, to this day, it is almost impossible to defeat rebels living in the many tribal areas bordering Pakistan and Afghanistan. As we went through the pass by train, I could picture many areas around craggy rocks and dead end valleys that an ambush could take place. We were told that whenever the train whistle sounded, we were to lie down on the floor, as tribesmen would likely shoot at the train. There were strategic locations where this often happened. Ben Laden if he is hiding, somewhere out there, has chosen an excellent hiding place to avoid his capture and punishment for his crimes.

There was a gathering of many of the Afghan tribes, who came to the city to attend a loya jirga (a grand assembly). I learned later that the tribes had assembled to discuss their neutrality with respect to the Second World War. Afghan tribes and neutrality, were like poles apart.

The Swiss, were neutral during the world wars, making cheese, chocolate and marking time by building watches. Those Afghan tribes would start a feud with each other at the drop of a turban. Out there it is still equivalent to the old Wild West. Instead of the cattle barons, there are poppy growers, instead of six-guns; there is an arsenal out there.

A large group of Afridi (one of the Afghan tribes located around the Khyber pass) was marching nearby. Perhaps, marching was not the right word to describe their movement. It certainly did not compare with the precision marching of the British and Indian army troops that I had witnessed. They were all armed to the teeth with several bullet holders strapped around their necks and shoulders. They were considered to be ferocious, crafty and treacherous and not to be trifled with.

A group of us, brothers and cousins, joined ranks, marching and, following behind them. They seemed to be amused by our efforts at keeping up with them. Our parents however were not amused, fearing that we may have been kidnaped, never to return home again.

There were and still are conflicts amongst tribal areas in the North West Frontier province of India which bordered Afghanistan. Warizistan, located in that area, was where my father was mentioned in dispatches and given campaign medals for his service with the army. It is also in Warizistan that Bin Laden is suspected to be currently hiding out.

I remember the huge interest in the sport of kite flying, which rivaled soccer and cricket, as a national sport. Kite flying was practiced all over India and Afghanistan. The kites were made with strong, coloured paper and thin slivers of bamboo were used to provide structural integrity. At times, there were scores of kites aloft. Against a clear blue sky, patterned and sporting all the colours of the rainbow, with some flying high and some low, it was a sight to behold.

I would often watch both young and old plying their skill. The main purpose was to maneuver ones kite to an attack position against any other kite flying in the vicinity. The kite flyers were just as skilled at maneuvering their prized possessions as a skilled fighter pilot maneuvering to shoot down his enemy. So when in an advantageous position, the kite would swoop down, making contact with the opponent's kite string and with a rapid pull cut it, or get cut. Note: The kite strings had been carefully prepared to be lethal, by soaking the string in a sticky substance mixed with powdered glass. Conquering an opponent gave much satisfaction, however getting friends to retrieve the downed kite, gave greater satisfaction. Cutting ones finger on the glass-coated string was a small penalty to pay. The biggest danger in the sport came from the fact that many fly their kites from the top of flat rooftops. The degree of concentration, while always looking up, caused many a tumble, which sometimes ending in tragedy.

This beloved sport was banished many years later during the Taliban occupation. What a terrible blow to the youth of the country when soccer was also banned and soccer stadiums were used to display executions. Thankfully, Afghanistan's new regime reinstated all sports and kites are up in the air again.

Another interesting and competitive game was "Gilly Dunda." The game was played with two wooden pieces. One, a notched stick, one inch in diameter and approximately two feet long was the "Dunda." The other, the "Gilly," was a smaller piece approximately four inches long, also one-inch in diameter and tapered at each end. A small circle was traced out on the ground using the Dunda and the Gilly was placed In the center of the circle. A player would enter this circle, using his Dunda, he would hit down on a tapered end of the gilly causing it to rise up in the air. He then struck at it, hoping the tapered wood would be sent a long distance away. Each person was given three chances. If he missed striking or was not satisfied with the distance achieved, he could try again. If the Gilly was caught in the air, the striker was out.

The distance traveled would be measured by counting end over end Dunda lengths, using notches marked on the Dunda to obtain a final fractional value to determine the winner. This was a very accurate measurement and prevented many arguments as to whose Gilly reached the furthest. I wonder if this ancient game could have been the forerunner of baseball or cricket. These accurate measuring skills were to be used later in my life when I spent many a day measuring and recording take off and landing distances on many runways by a variety of means.

Most of my life in India was spent in Calcutta, Bombay, Poona and Old and New Delhi. In Poona, our mailman brought in the mail by camel and often took us for rides on top of his delivery vehicle. Sitting atop that tall mammal, I felt like the king of the hill, with the ground far below. That strange longitudinal fore and aft motion remains with me and now reminds me of an equivalent sensation when encountering a water handling problem on a float plane or sea plane. This problem, referred to as porpoising, is encountered when an aircraft dips in and out of the water in an unstable and unacceptable manner.

My father, a captain in the Indian army, was posted in Warizistan, Baghdad and Basra. He was mentioned in dispatches for his outstanding work. He didn't talk much about his experiences during the war, but when he did, we could listen to his stories for hours. I remember him telling me about the day he married my mother. At the reception after the wedding, he took off her silver shoe, poured Champaign into it and proposed a toast to her.

My father was a generous man. He not only supported his mother who had been a widow for as long as I could remember, but also sent two of his brothers to University to study teaching. One uncle taught in

Burma, the other in India. I was taught geography by an uncle who was employed for a short while, as a teacher in St Xavier's Jesuit school in Calcutta.

My mother who lived on a rubber plantation in Burma, now Myanmar, met my father in India when her parents sold the rubber plantation and bought a tea estate in the Nilgiri hills in South India. I had an elder sister Cecily, who was one year and a day older than I, three younger brothers, Peter, Winston and David and a younger sister Margaret.

Mother Teresa who was then Headmistress at Loreto Convent in Calcutta had taught Cecily. I was told that she also taught me when I was in kindergarten there, but unfortunately I do not recollect that memory. When my father retired from the army, he secured a position as Chief Accountant and Auditor, for many of the Indian Railways. We often traveled with him. The railway networks that my father was involved with, gave us an opportunity during school holidays to travel all over that vast country in luxury. A large saloon railway car, when not required by the governor of India was made available. This car was attached to a train, and then, detached and parked on a rail siding in the many cities, where he conducted an audit at various railway stations.

The railway companies he worked for encompassed most of India for example, the Great Indian Peninsular (G.I.P.) Railway, the Bengal and Nagpur Railway (B.N.R.), the East Indian Railway (E.I.R.), the North West Frontier Railway (N.W.F.R.), South Indian Railway (S.I.R.), the Mysore State Railway (MSM), the Great Northern Railway (GNR), and perhaps a few others I may have forgotten.

There are four different gauges (width of railway tracks) in India, broad, intermediate, narrow and extra narrow. The width between rails varied from just over five feet, to two feet in mountain regions. The Governor's saloon we sometimes traveled in operated on wide gauge tracts. The saloon car was painted a gleaming white with gold lettering and prominently displayed was the royal crown emblem. It consisted of three sleeping compartments, each sleeping four, a lounge area, a kitchen as well as quarters for the contingent of servants that came with the car. We were made to feel like Lord and Lady Muck. I wonder if Lord Louis Montbatten used that saloon as the last Viceroy of India.

Other gauge railways had V.I.P. saloon cars, but none quite as grand as the Governor's. Traveling in style and comfort was not too shabby and my father took his family on many of his auditing trips throughout the length and breadth of India. My recollection of the names of many

Indian railways pays testament to those travels.

Often, upon our arrival at our destination, my father and mother would be showered with garlands as people in India had a tradition of plying garlands on people with perceived superior rank. On arrival at many locations, my father and mothers would be festooned with highly scented flower garlands, Mountain railways always operated on narrow gauge tracts, as it was easier to carve out a narrow track in difficult terrain. Before encountering a steep gradient, the train would stop and two brave railway employees were dispatched to the front of the engine. Here, they were perched precariously on the cowcatcher, which is a plough like devices used to scoop up objects such as cows, other animals and fallen rocks, to prevent derailment. They were each provided with a large bucket of sand. As the train started up a steep slope, they would pour sand on the left and right rail line. This increased friction and provided better wheel grip. Many years later, in North Bay, Canada, I would be engaged in a program to measure the friction on an icy runway and show the improvement in friction when the runway surface was sanded down Thus we visited many of the Indian Provinces, which then included Pakistan and Bangladesh and marveled at the striking variety of architecture, language, culture, dress and ethnic features. There was a vast contrast between Hindu cities and Islamic or Muslim cities.

The Hindu holy city of Benares now Varanasi, contained many temples and cremation fires. Here, ashes of the dead would be sent into the holy Ganges River. Disposing human ashes from an airplane in accordance with the deceased wishes will be a subject, later in this book. In striking contrast, was an Islamic city like Lahore, with its many mosques and with constant sounds of religious leaders calls to prayer. This variety was even evident within a city. The most stunning contrast was between Old and New Delhi. It was ancient then suddenly modern after a short tonga (horse buggy) ride.

We made several visits to the Taj Mahal in Agra. The Taj Mahal is rightly named one of the world's wonders and a fitting tribute of the love of an emperor for his wife. I have been there several times and the sheer beauty of that large gleaming white marble monument, framed by four tall minarets, did not deteriorate with familiarity. A sight to remember is the upside down reflection in the long pool leading to the monument. The reflecting pool is aligned with tall narrow trees and colourful gardens.

We had an ayah (nanny) who was with us for many years and became part of our family. She sang us a lullaby at night and I remember it well. This is that chant, which was repeated time and time again.

Ninny baba ninny "Sleep child sleep"
Mucken roti chinny "Butter bread sugar"
Mucken roti hogia "Butter bread is finished"
Humra baby sogia "My baby has gone to sleep"

Around the age of six, though I loved her dearly, I thought I was too big to hold my ayah's hand. That was before I decided to climb the steep spiral staircase to reach the top of all four minarets of the Taj. On my way down from the third minaret, I suddenly came across a Sadhu (holy man). I was running down those steep steps and just near the bottom I collided with the sadhu. I remember staring into his dark piercing eyes. He was completely naked. His long gray-mattered hair matching his long beard and he was covered in ashes. As I collided with him, he yelled, I yelled and we both tumbled down the few remaining steps. I was quite shook up and ran straight into the arms of my ayah and would not let go of her hand for all the tea in India, much to the amusement of the rest of the family. Still to this day, when I see a picture of the Taj Mahal, which is invariably taken from the front of that beautiful reflecting pool, I look at the rear right minaret and remember my contact with the holy man. I also recall looking over a stone rampart and down below was a large collection of crocodiles basking in the sun. Across the river and not too far away was a group of washerwomen busily washing clothes on the riverbank. I could hear the slapping of cloth as they traditionally beat the clothing against a flat rock, seemingly oblivious to the hazard across the river.

The contrast between riches and poverty was evident in the India I knew. Poverty is a relative word. Today, whenever I hear poverty mentioned in the western world, I contrast it with the real poverty of the undeveloped countries and picture people dying on the streets with vultures eating their rotted bodies. There were many beggars, some with advanced leprosy, manifested by grotesque sores and missing extremities. Others had gross deformities, some with small heads as big as a fist, some with two heads, one I saw with teeth growing out of his neck. When I was told that many of the blind and limbless beggars were deliberately maimed and sent out to beg, I could not believe it. As time marched on I have come to know that such things did happen.

Lunchtime at St Xavier's College in Calcutta provided a clear reminder of the poverty prevailing. The students left over lunches were tossed into large garbage bins, which were then taken to the street by a side gate. The eager waiting crowd of the poor and hungry then rummaged through for scraps, fighting over them, like bears and raccoons at a Northern Canadian dump.

Real poverty is hard for some to imagine. Mother Teresa saw it and acted to help the hopeless. She eventually left the teaching convent and obtained permission from the Vatican to start the Missionaries of Charity, devoting her life to lepers, the poor and homeless.

There were many ways of keeping cool during the hot summers in India. The obvious, was to head for the hills. With temperature dropping two deg C's for each, 1000 feet increase in the altitude, the foothills of the Himalayas and the tea plantations in the Nilgiri Hills, provided a great escape from the searing heat down in the plains. Ice was always available in the bazaars of the major cities. Huge blocks of ice, manufactured in many ice factories, were placed in bags of sawdust to provide insulation. Ice picks were used to cut out hand size portions for purchase. Sucking on a chunk of ice was quite a treat. Ice, when mixed with boiled green mangoes, water, sugar and milk, created a delicious refreshing drink, called mango fool, which had a taste sensation not to be forgotten.

Filtered cold water was produced the following way: - Tiers of semi porous water goblets were located in a wooden frame. On top, was a goblet filled with sand with a hole at the bottom, a similar goblet was below, filled with charcoal and below that one, was the drinking water goblet. When water was poured into the top container, it would dribble through to the drinking goblet. Evaporation due to the water seeping through the porous earthenware material would keep the water cool. I can still hear the glug, glug, glug, as clear cool water was poured into a glass, which would soon have beads of condensation running down on the outside

Another application of the cooling effect of evaporation was the "kus kus tattie." This was a one-inch thick blind, woven from root fibers and was rolled up over open windows. When rolled down, buckets of water would be thrown onto the kus kus tatties. The roots absorbed the water, which evaporated and reduced the room temperature significantly. Cooling, due to evaporation, played a significant role during an episode in my future career (more on that later).

CHAPTER 2

Just outside Ootacamand or Ooty for short, in the Niligri Hills in South India, was the location of my grandparent's tea estate, which also had a section of coffee trees. The plantation bungalow was located on top of a small hill overlooking the estate. A wide veranda surrounded the home, which sported a beautiful garden, tended by some of the plantation workers. Behind was a nursery, where young tea seedlings were grown. When reaching approximately 12 in high, they would be transplanted onto hilly slopes, with many shade trees to protect them from the sun. There was an animal enclosure near the house that housed chickens, ducks, cows and a donkey. There were also vegetable gardens and many fruit trees nearby.

Often, jackals and occasionally, a leopard would try to get into the chicken coupe. Loud squawks would alert my grandfather to collect his shotgun and scare the raiders off. The estate was pretty well selfcontained.

Milk and butter, came from a small herd of dairy cows, with Molly and Topsy, our favorite animals. The chicken coop provided fresh eggs and sad to say fresh chicken meat. Apple, peach, plum trees, a fig tree and berry patch, provided fruit and homemade jam. Vegetables, from a garden patch created balanced meals. Staples such as large sacks of flour, rice and lentils and barrels of salt beef, negated many long trips into Ooty for supplies.

Once on return from a trip to Ooty, my brother, sister and I were in the open rumble seat at the back of the car when we came across a rare black panther. We passed within a few feet of this beautiful jet black animal, which snarled with big gleaming white teeth as we went by.

My uncle was put in charge of the running the estate as my grandfather was getting old and left the day to day operation of the estate, to his son, my uncle. Early morning he just blew a whistle, which would cause the workers to start picking young tea leaves, and tossing them into a big basket, which was resting on their backs supported by a wide headband around the forehead. The next two whistles were to give them a lunch brake and send them back to work. The final whistle signaled the end of the day's work. My first impression of my uncle's running of the estate was that it was a piece of cake. I know now that there is more to controlling any operation than meets the eye, especially the eye of a young boy. The tea pickers, who were all women, would get their baskets of tea leaves weighed at the end of each days work. Some baskets contained choice pickings of the finest quality tea. These were the fresh buds with two leaves immediately below. A higher price was paid for these choice baskets.

Life on the tea plantation convinced me to study agriculture and become a tea planter.

While visiting here, we spent many exciting times exploring the jungle. Although it was dense, there was a narrow mossy trail and we got to know it well. The jungle had a distinctive tone, buzzing with the sound of insects, mainly crickets as well as colourful birds and chattering monkeys. Tree trunks were laden with moss out of which many orchids grew. There were beautiful coloured butterflies, dragonflies and mosquitoes too. At a particular location on a large moss-covered boulder we often came across marble insects. These were about two inches long and when disturbed, would roll up into a perfectly formed ball and roll down, presumably as a method of self-preservation. We would wait patently to see them unravel and crawl away on the multitude of legs they possessed. Occasionally we encountered an army of leaf carrying ants forming their own trail across our path. We never felt that we would face any danger, until one day, we noticed my sister standing still as if in a daze. We then noted a cobra with its hood open, swaying from side to side. The deadly snake was about four feet in front of her. We pulled her away before it struck and headed home much relieved that the snake did not follow and attack us. Later she said that she felt that she was in a hypnotic trance. Those swaying beady eyes could have been the cause.

Near by, was a village that was inhabited by a tribe called the Todas. They were a primitive tribe. Later I learned, by reading the National Geographic Magazine, that the Todas were one of the oldest living tribe in existence. A name I should never forget was Gidgee Gunda, who became a legend in that area. He was a foreman on the estate and was the one that got the tea pickers to respond to the whistle. My recollection of this man is extremely vague, so I will describe him, the way I remember him, an unassuming heroic middle-aged man with a greying beard and a stocky build. Gidgee Gunda and his crew were engaged in pruning the tea bushes. This was a regular function for the men, in order to prevent the tea bushes turning into trees. It was during this process that a tiger appeared and seemed to be stalking some workers, who began running with the tiger in pursuit. Gidgee Gunda ran toward the tiger shouting loudly and throwing stones. The tiger ran away. Thus, legends are made.

My sister Cecily, and brothers, Peter and Winston and I, would often watch birds nesting in tea bushes, their eggs turning into little birds. Paying regular visits to these nests, one day we saw a large snake in a bush with the nest empty. The baby birds were gone. We promptly killed the snake with sticks and stones, and started a vendetta against snakes. Armed with bows and arrows used by the local tribe, we began our snake hunt. We killed many snakes, about five, sometimes ten a day. What a stupid and dangerous game we played. The arrows were much more lethal than the small pellets we used when shooting with a Daisy air gun. We became quite accurate when engaging the snakes from a few feet away. Often we would practice by firing a long distance aiming at a tree trunk. Those arrows flew straight and true. Many years later I would be undertaking rocket-firing trials and watch those rockets like those arrows, also flying straight and true.

I recall the day I went swimming in a small lake, while camping with the boy scouts, only to later see hunters with express rifles, shooting crocodiles in the very same lake. On growing up I came to realize that the Gods must have been smiling on me because I believe I led a charmed life.

Thanks to a colleague of my father's, who was a famous Shikar (hunter), we went hunting leopards and wild boars in Gobardunga, which is in Bengal Province. It was another memorable experience. My sister was very upset when she was not allowed to go on this trip. She would never understand why being a girl prevented her from getting this experience. Volunteers from a nearby village were gathered, to act as beaters. They were to be paid by sharing in the spoils. Hundreds turned up, armed with sticks, drums and blow horns fashioned from cow horns. I was fifteen years old at the time and pretty excited. Armed with a heavy double barrel shot gun, each barrel loaded with a rotex shell, which rotated when fired, making a loud whistling sound. I had never fired this size of a gun before. It was quite an effort to lift it and aim. I was told there would be a hefty kickback when fired. The beaters hacked their way through the jungle, shouting loudly and banging sticks and drums and tooting their horns. I had been positioned below a tree, which could be climbed in an emergency. I was sure a leopard could shin up that tree a lot faster than I could. In front of me was a small clearing. The excitement built, as the beaters got closer and closer. My heart was racing. Perhaps I should admit that I was a bit scared. Three boars suddenly appeared. I had been told they were more dangerous than tigers. These three had large curled tusks and were ambling along slowly. I fired my shot and heard a squeal, as my target ran away. Seconds later I heard another shot. The shikar had killed the wounded boar, which had been crippled by my bullet that struck its hind leg.

A second beat took place a couple of hours later, this time, as they were heading toward us, the shouting intensified. The shikar paid me a quick visit, to tell me that the beaters were calling out, that they had spotted a leopard and also told me that I should take a shot as soon as I saw yellow. As the beaters were almost up to me, I spotted something yellow and cocked my gun and aimed at the yellow, when suddenly I saw a beater with a yellow turban. It could easily have been manslaughter. The Gods did not want me to have that on my conscience. Hunting laws today, ensure that hunters and beaters must wear florescent clothing. I often wonder how many beaters had been shot during those old hunting days.

A day later the leg I brought home, was skinned and we had pork curry that night. I took the skin complete with my bullet hole, scraped it, salted and dried it for a while in the sun, hoping to keep it as a souvenir. The smell did not go away and my mother told me to throw it out, which I did, on top of an armoire, in my bedroom. Later the sweeper complained to my mother that he kept finding maggots in my room. An intensive search revealed a maggot infested skin on top of that armoire. The skin I was so proud of was discarded. Lesson, always listen to your mother.

My father was invited to stay at the Raja of Talghar's palace somewhere in Bengal. I cannot remember exactly where it was, or how my father got to know him. I do remember that the young Raja had a small kingdom, which also included a coalmine that no doubt added to his riches. Perhaps my father was negotiating a coal contract for the railways. I will never forget the immense living room where we were entertained. On top of a thick red carpet, tiger and leopard skins were strewn about, each with heads with mouths wide open, displaying large yellow teeth. A large black ebony table in the center of the room, had legs comprising stuffed elephants' feet. Footstools made from elephants feet, were strung around the room. I sat on one. The walls were adorned with many hunting pictures, most with tigers, featuring the Raja and his friends standing or sitting proudly beside their kills. Today the Bengal tiger is near extinction, and I wonder how much of this tragedy was caused by our charming host. We were taken to see his private zoo. Amongst many species of caged animals were Bengal tigers. I became a witness to an amazing sight. The Raja sent one of his servants into a tiger's cage. Here I saw a display better than I have ever seen in a circus featuring animal trainers. There were two large tigers in the cage. The servant opened the cage door showing no fear whatsoever. He then proceeded to wrestle with the animals. All three were soon rolling about on the floor amidst snarls and occasional roars. Man and beasts were playfully slapping at each other. Before exiting the cage, the man opened the mouth of one of the tigers and placed his head in it.

I can only assume that those tigers were obtained as cubs after the mother had been shot. The next day we were taken to the Raja's coal mine where I jumped at the chance when asked if I would like to go down in the mine. It was a long way down in a small cage, extremely damp and cool with water dripping down the mineshaft. A long rumbling ride on a wheeled bucket took us to the coalface. We then reversed the process and returned to the surface.

Many years later, after suffering the "bends" during a test flight, I was placed in a pressure chamber and taken down below sea level and kept there for many hours to alleviate the malady. It was just like going down in that mineshaft, deep into the bowels of the earth.

We often visited my grandmother, who lived in Vizagapatan, (now called Vishakhapatnam). It was a seaside town on the east coast of India. My grandmother's house was quite a distance from the railway station and a short distance to the beach. When we arrived, transportation to her house was by bandy cart. This was a box like compartment on solid wheels, with a thatched roof, pulled by an ox. Direction control of the vehicle was by tugging on a rope tied through a hole drilled in the animal's nose. The cart driver would often beat the ox to increase speed. As the cow was such a sacred animal in India, I wondered why such cruelty? The solid wheels and a rough road contributed to a very bumpy ride.

I have wonderful memories of endless summer holidays at grannies' house. We were often at the beach swimming and watching fishermen launch their crude catamarans through the surf and out to sea, maybe returning several days later. Sometimes I would help pull in the nets laded with fish of all kinds, including hammerhead sharks.

Grannies' house was large, with lots of flowerpots strewn everywhere. There was no plumbing, so every day, water women brought in large earth containers filled with water, balanced on their heads and deposited the water into a metal tank located in a large washroom. This washroom was often referred to as the throne room. Around the throne room were several ornate wooden chairs with comfortable arms and a round hole in the seat area. Below the holes were ceramic pots. On a shelf were a variety of lids, some made of carved sandalwood and others of porcelain. It was a rule easily learnt, not to use a throne with a lid on the hole. Twice a day, sweepers (the lowest of all the casts) would remove the capped pots, and return them fresh and clean. Where they deposited the contents, I have no idea.

Vizagapatam or Vizag for short was a seaside town with a large deepwater harbor, from which manganese and coal were shipped all over the world. A cousin of mine was harbor- master there.

The movie theater in this idyllic holiday place, had a corrugated tin roof, held up by wooden posts, allowing the cool sea breezes to pass through, thus providing a poor man's air conditioning. The seating arrangements here reflected the class culture always present due to the Hindu cast system as well as the British class system. The front row seating was a sand patch, then benches, then hard chairs, then chairs with seat pads followed by plush arm chairs and finally box enclosures reserved for the big wigs. Often the movie projector would break down and the "chokra" boys (young teenagers) would rush out and shower the tin roof with stones and small rocks to announce their displeasure. This was accompanied with boos from the rest, except perhaps from the occupants of the box seats. It was here, that I saw that movie about a test pilot, which I refer to later when describing high-speed flight.

It was in Vizag that we witnessed the first air raid by the Japanese in India. We were on the train at the time about to leave for our home in Calcutta. I looked up and saw a formation of aircraft flying low and was bragging that I knew they were Kitty Hawks. When tell tail puffs of smoke suddenly surrounded them, accompanied by the distinctive pom, pom, pom, sound of ack- ack (anti aircraft) guns, I realized that my aircraft type recognition needed some improvement. The aircraft were Japanese Zero fighter/bombers. The gunfire was from, naval ships based in the harbor.

We ran prudently to the railway station waiting room with bombs falling very close by and parked ourselves under a heavy table. The "punka wallah," whose job it was to keep passengers in the waiting room cool by pulling on a rope, like a bell ringer, moving a large mat about two feet wide and swinging it across the ceiling to create a breeze. He continued in his task throughout all the commotion, determined to keep us cool. We told him to quit his job and join us under the table. I kept running out from under that table to catch a glimpse of the enemy aircraft much to my parent's alarm. A bomb landing nearby caused me to scamper back to the apparent safety under that large table. About five minutes after the attack started, the air raid siren sounded. An hour later, the long continuous sound of the all clear sounded. We headed back toward my Grandmothers, house to check up on her. On the way, a formation of three aircraft flew overhead. People around started to panic thinking the Japanese were back. I knew without doubt, this time, that they Westland Lysanders, a small spotter aircraft with a very distinctive wing shape, it had a reputation for taking off and landing from very small airstrips. They were fitted with small bomb racks and would be no-match for the Zeros. As they headed out to sea, apparently in search of the enemy aircraft carrier, I wondered if they would ever return.

Many years later I worked on the development of a short take off and landing aircraft with a short field performance not unlike that of the Lysander.

We often explored the back streets and villages around Vizag, where I got an insight into the local living conditions. Open sewers were common, you could tell by the smell. Cow dung was used for a variety of purposes. Cow dung when mixed with water to form a thin paste, was used as a brown wash to paint the floor, outside walls and front doorstep. When dry it had virtually no smell, if anything the smell had a slight pleasant odor. Spread over the doorstep, it provided a shield against dust. The doorsteps were adorned with elaborate designs, made from white and coloured chalk. It appeared that there was a competition between neighbors as to who had the best artistic display. Pancakes of cow dung were slapped on the outside walls, which fell off when dry and provided fuel for cooking. Often we would see local boys and girls following a cow, and scrambling to grab hot fresh fuel. They placed their winning in a basket to take home to their mother. I remember well, a young girl only about two years old, running behind a cow and collecting a handful, as it dropped steaming in her tiny hands. She was smiling as she handed over this precious gift to her mother who hugged her in a show of appreciation. Other animal droppings were also used for cooking fuel, such as horses, donkeys and camels.

The milkman brought his milk generator straight to the back door. He squeezed fresh milk straight out of his cow, which was usually accompanied by its calf. If the calf died, it was skinned and stuffed with straw and placed next to the cow to encourage milk production. Occasionally the mother would lick the dead carcass, in a sad show of affection.

With current highlighting of the ecology and global warming, the lifestyle of the so called underdeveloped countries did very little in those days to contribute to the problem. Food vendors served food on banana leaves and tea was poured into thin clay cups, which after the contents were drunk, were then tossed back to the ground from whence they came. Transportation was mainly by rail. Trains were often overcrowded with people clinging to handrails on the outside of the train and also perched precariously on the roof. Animals providing the power to propel vehicles and humans pulling rickshaws and carts did not cause pollution. Irrigation was by ox driven water wheels and sea saw scoops, which raised water from one paddy field to the next. The scoops were made from hollowed out coconut trees and operated by manpower. Another irrigation method entailed a man or woman who would walk back and forth away from a pivot point, thus filling the scoop and depositing water to a higher level. Yet another method was two persons standing opposite each other, dipping a large bucket made of animal skin into the water. Both would then haul back on two ropes tied to the bucket and deposit water from one level to the other.

Road building was another example of the use of manpower. Large road gangs would clear out brush and rocks in the proposed roadway and then in military precision, would march, singing a rhythmic chant. They marched about twenty abreast and three or four rows deep. Each carrying a pole attached to an approximate fifty-pound flat hammerhead.

They would raise and lower the poles following the rhythm and advance slowly up the road, pounding it into shape.

The simple approach often portrayed in India, registered, an impression, allowing me later to seek out simple solutions to complex problems.

Experience, dealing with pariah dogs also surprisingly played an important part in my future career. I will explain. Pariah dogs were vicious looking mongrels, they were outcasts, and apparently owned by no one. They flashed their teeth, which were white with a tinge of pond scum yellow. They snarled and growled revealing pink jowls dripping with slimly spittle, at every encounter with a stranger. Often they would gather in packs, posing a serious threat. When encountering these beasts, any show of fear and a panic retreat, would unleash an attack. One had to remain cool and calm with every encounter. The dogs were just as fearful of you as you were to them. If you showed aggression toward them, they would skulk away with their tails between their legs. I am sure that learning to conquer the fear of these dogs enabled me to remain calm when encountering the inevitable confrontation with unexpected gremlins that created dangerous situations in my future flight testing career.

A sad experience with a pack of pariah dogs occurred when holidaying with my family in Vizag. We took possession of a stray puppy dog. It was a happy, friendly dog as most pups are. One day the pup was morose and did not show much sparkle. The pup was outside the house when it was attacked by a pack of dogs and was ripped apart, dying instantly. We all felt that the dog had a premonition of what was to come and hence its prior sad mood.

My father was transferred to several northern railways so we moved from our home in Calcutta to a large apartment in New Delhi and soon after to some civil servant's bungalow on the outskirts of the city and finally to a very large home on the outskirts of Old Delhi. I was 14 years old at the time.

New Delhi was a modern city, easy to navigate with broad streets radiating from the center. There were two ring roads, one circulating an inner circle of buildings and the other an outer circle of buildings. The inner circle known as Connaught Place was where the apartment was located, right in the center of the city. It was a large apartment with high ceilings and marble floors. At ground level were shops, restaurants and movie theaters. We did not stay here long, but long enough to give me a taste of city life.

Our next move was a temporary one as we awaited the completion of a large dwelling located in Old Delhi. We moved into a vacant civil servant home. It was summer and it was hot. The streets were mostly deserted, as almost all the residents had headed for the hills. All senior officials had two homes supplied by the government of India. When summer was nigh, an army of civil servants would move up to homes in the mountains such as Darjeeling and Simla. Here they would continue their work in duplicate offices. Naturally, all their servants went with them.

One day, after a bicycle race with friends, around deserted streets, I suffered a heat stroke, was delirious and had a body temperature of 106deg F. With lots of cold towels and ice, I survived. Years later, I heard stories of near death experiences and a common theme in those stories was an experience that I was able to recollect. This was floating about ten feet above my sick bed and watching a doctor, a nurse and my family surrounding the bed. There was a bright white light I remembered and then slowly I fell back into my body. Chalk another one to the Gods. No wonder those civil servants head for the hills in the summer months. After I suffered that heat stroke, the family never went out without solar topees (pith helmets) thus keeping their heads protected from the hot sun.

When summer was over, we moved to Old Delhi. The house was three stories high and on the roof, were three empty separate, self-contained apartments. There were twenty-three rooms not counting the apartments above. Across a beautiful garden courtyard, were the servant's quarters. We played great games of hide and seek in the many empty rooms. Exploring the rooms, we came across a locked door, that once opened, led to our discovery of an ornate temple complete with statues of many Hindu Gods. The statues were painted a bright indigo blue with plenty of gold and silver. There was an elephant God, a God with many arms and even male and female Gods doing naughty things

Each morning, two tongas (horse drawn buggies) would take us to St Colombo's Irish Christian Brothers School in New Delhi. Sitting in a tonga going from Old to New Delhi demonstrated the vast contrast between the two adjacent cities. Old Delhi had crowded narrow streets with lots of slum areas and smells due to poor sanitation. New Delhi was a clean modern city. We were in Delhi for less than two years when my father was transferred back to Calcutta.

CHAPTER 3

Throughout our days in India, ever since I could remember, we owned a fox Terrier called Dot. When Dot died of old age, a journalist friend of the family, who traveled all over the world, gave us a black and white puppy and were told it was a "Fox Terrier." This pup, which happened to have great big paws, was named Jock. Well! Jock kept on growing, to such an extent that he was referred to, as the family cow and was as big as a Great Dane. Jock was a good companion, a threatening, but docile, watchdog. We took him on a visit to Naini Tal, where my uncle, the doctor had left Peshawar, retired from the army and became a surgeon at the local hospital.

Naini Tal is a hill station in Northern India at the foothills of the Himalayas. As you arrive there in a bus, through a mountain gap, you are suddenly confronted with a beautiful lake, surrounded by pine-covered hills. There is an area called the flats at one end of the lake, which was formed when a devastating mudslide engulfed the area, causing the death of one hundred and fifty-one persons. The flats are now used as a recreation area with soccer and field hockey grounds. Here you could hire horses and follow many trails into the hills. Nearby is a boat club, where we rented boats and joined our many cousins in days out on the lake. I will never forget the day we rented horses for a trek into the hills and were directed to a path, which soon became narrow with steep cliffs sometimes on both sides. I was in the lead at the time, when my horse suddenly panicked; it had spotted a dead monkey, reared up, turned around and bolted past the other horses. How we all did not end up in a bloody pile at the bottom of a steep cliff, I will never know. Once again, my thanks go to the Gods.

Our dog Jock had got into a fight with some large monkeys. A few bruises but he seemed O.K. After we left Naini Tal, during the two-day train trip home, the dog appeared listless and bad tempered. On arrival at the Calcutta station, Jock went berserk and bit my brother Peter, myself as well as a couple of coolies who were carrying our luggage. The next day, Jock was foaming at the mouth and running about aimlessly.

It was rabies, probably from those monkeys. Jock was put down, and the search began to find the coolies that had been bitten. The stationmaster eventually found them, and we took them with us, every day for fourteen days, to the hospital. Each day we were injected in the stomach, with a very large syringe containing anti rabies medication. Without these injections we were told that we would all be doomed to suffer an agonizing death. Each day we lined up standing erect and watched our stomachs swell with the large amount of serum injected. It seamed to grow more painful each time. It was a great relief for all of us when the fourteen days were over.

Back home in Calcutta, a few weeks later I heard my mother at her dresser. My back was to her, and I started into a one way conversation with her. Getting no response, I turned around to find not my mother, but a big spider monkey, powder strewn all over, including its face. I shouted. It then ran all over the place and jumped out of a window onto a tree, perhaps smelling like a rose. It was a really big jump to reach that tree and perhaps worthy of an Olympic medal. Monkeys crossed my path once again many years later, during flight tests in the Persian Gulf Island of Bahrain. This episode will be described later. Soccer, field hockey and cricket marked the seasons. Cricket was not my game, there was too much sitting around waiting your turn at bat, in any case, I was a lousy cricket player.

Between sports and studies, school was over before I knew it; I had taken Latin and Hindi as my foreign language subjects. Greek and Sanskrit were also offered. I believe Latin, even though criticized as a dead language gave me a good grounding into logical thinking. Later in life when conducting aircraft tests in Peru, I could understand some of the conversation in Spanish without ever studying the language. As for Hindi, I already knew the "Pidgin English" form of the language, known as Hindustani, which was used to converse with the servants. My limited knowledge of Hindi came in useful only very recently. I was having lunch with friends in Huntsville Ontario, Canada. On the next table was a young couple. She was wearing a blouse with a word printed in Hindi, splashed all over that blouse. I surprised myself that six decades after leaving school I was able to read those letters, all the more difficult as she would not sit still. Written on her blouse, when translated, was Sitaram. Excited at my memory recall, I asked her, if she knew what was written all over her blouse. She replied that she did not even know what language it was. I told her that many years ago I studied Hindi and that Sitaram was a Hindu Goddess, the wife of the God Ram. Her partner was pleased and said he always knew that she was a goddess. He, no doubt, was attempting to gain brownie points.

At the age of 16, I completed the Senior Cambridge School leaving Certificate. University came next. I studied Science at Calcutta University and later became a member of the University Squadron and learnt to fly in a Tiger Moth aircraft, which is a two-seat biplane trainer, built by the de Havilland Aircraft Company, a company that I would work for many years later. I remember my first flight in the front seat of that open-air cockpit; I could not wipe the smile off my face for a long while after landing.

Calcutta was bombed several times by the Japanese. It is a strange coincidence that the only two places in India, bombed by the Japanese, was where our family happened to be. When the air raid sirens sounded we went to a "safe" location under the stairway. Here we could hear the anti aircraft guns firing at the enemy aircraft.

The R.A.F. had two squadrons of fighter aircraft one operated spitfires, the other hurricanes. The airstrip they were using was actually a main street in the heart of Calcutta called Chrowingee. The street was cordoned off and converted to an airstrip. Being located not far from where we lived in Calcutta, I often spent time there, once again watching fighter aircraft taking off and landing, as I had done years before in Peshawar.

One of the fighter pilots operating from this the city strip, whose name I still remember was Flight Sergeant Pring. He flew a Hawker Hurricane and was awarded the Distinguished Flying Cross for shooting down three Japanese zero fighter/bombers right over the city, accomplishing this feat in only four minutes. I was then determined to join the Royal Air Force.

The achievement of building an airfield in the heart of a city during wartime does not escape me, as many years later I was to be involved in the development of short take off and landing aircraft for city center operation

A favorite comic book hero of mine was Rockfist Rorgan of the R.A.F. He together with Mad Max Carew, another dashing pilot, was featured in the Champion comic book. Every week, stories of their exploits in the air, enthralled me, and I could hardly wait for the next week's issue.

As I was passing an R.A.F. recruiting station, I got a sudden urge, walked in and filled out aircrew application papers. I was told that the first R.A.F. flight-training establishment in India had recently been opened and was conveniently located about fifty miles near my home in Calcutta. Soon after my application the war with Japan was over, the flight-training unit near my home was disbanded, all personnel were returned to the U.K. and my dreams of emulating the exploits of Rockfist Rorgan and Mad Max Carew were dashed. Instead, I continued with my studies unaware that soon I was about to witness, terrible scenes that eventually led to India's independence.

The desire for independence, led to massive demonstrations, leading to police intervention and riots. Soon, the protests against British occupation, changed to ethnic cleansing. With partition in mind, Hindus and Muslims were slaughtering each other to gain the upper hand in their areas.

About that time, after a game of rugby, I was on my way home with an Armenian friend; both of us were battered and bruised, from the game. We were suddenly, attacked by a mob, who threatened throw us onto a big bonfire that had been set on the street. A distinguished looking Sikh gentleman, who radiated authority, assured the crowd that we were Armenian and not British. He saved us from a near death experience.

Ever since then, I hold a lot of respect for Sikhs and thank God that I became an honorary Armenian that day.

Our cook, who had been with us all the nineteen years that I lived in India and who traveled all over with us, brought home, the horror of those troubled times. He was a Hindu and went to do the shopping in an area he had always shopped, which happened to be a Muslim area.

He was from the south of India and was as black as ebony. Returning from this shopping trip, he was shaking with fear, and the whites of his eyes were shining in sharp contrast to his black skin. He informed us that the meat hanging on hooks, which he was about to buy was Hindu flesh. It was sufficient warning never to go to that area again.

A neighbor of ours, who was chief of the Calcutta police, and happened to be a Muslim, answered a knock on his front door to discover a suitcase there. He opened it to find a chopped up body of his son, who went to school with us.

One time I was walking down a quiet street near home, I noticed some young kids who seemed to be playing and dancing around an old man. I can still hear their laughter. Next, I saw him fall to the pavement bleeding profusely. They were not playing, but stabbing him to death.

The first aid I had been taught while in the Boy Scouts and later in the King Scouts, did him no good, he was dead when I got to him. A crowd soon gathered and I hurried back home.

The famed Gurkha regiment with their British officers had been called into a rare action to help quell some of the rioting. These small but tough soldiers had a reputation that all feared. Armed with kukries (large angular shaped knives) and batons I once saw them charge into an unruly mob, scattering them and grabbing some of their ringleaders.

The British forces were ordered to remain neutral and unfortunately did not prevent the mass slaughter that was taking place. It is said that more people were killed during those riots than all the British soldiers killed in the First and Second World War, put together. The stench of truckloads of bodies and the sight of flocks of vultures consuming bodies in the streets will remain, thankfully, as a distant memory.

The University, in Calcutta was shut down due to all the rioting and I continued my University studies at St Edmunds College in Shillong, Assam, located 6000 feet above sea level at the foothills of the Himalayas. Two of my brothers were in boarding school there. Later, they went to St Joseph's College in Naini Tall, which was further North and also in the foothills where my uncle was now a surgeon practicing in the local hospital. Naini Tall was a picture post card town, centered on a perfectly rounded and very deep lake that had once been the crater of a volcano. A number of my cousins lived there and it was a great place to visit.

My elder sister, Cecily, was sent to England to study law at London University. The younger ones stayed at home in Calcutta but in an area well away from trouble spots.

The College in Shillong was run by Christian Brothers and contained grades from Kindergarten to University.

To reach Shillong, one had to travel by train from Calcutta to Dacca, followed by a long river paddle boat trip on the Brahmaputra river to Gouhati and then a long winding bus journey up through the Himalayan foot hills through dense jungles inhabited by tigers and elephants. The journey took three and sometimes four days. During the paddle-boat journeys I often noticed large Shorts Empire flying boats skimming fast, on or just a few feet above the river. Later, I found out that the airline company operating routes in that area had their operating license withdrawn. No worry, they continued by taxing at high speeds and operating as a high speed boat service, carried many passengers and goods to a number of towns and villages up and down the river. I understood that by not operating as an airline, they were able to overload the flying boat and even operated the "aircraft" with mechanics and not qualified pilots. Perhaps it was a profitable but risky operation.

Wouldn't you know it, many years later I worked at Short Brothers and Harland, the aircraft company that built Empire flying boats. There was a fuselage of a Solent (similar to the Empire flying boat) parked in a corner of the airfield. I examined the fuselage carefully and seriously considered converting it to a home. My offer to take the derelict fuselage off the company's hands was not accepted. I guess they would get a lot more selling it for scrap. Many years later I acquired a fuselage of a de Havilland Dash 7 aircraft and used it as a weekend cottage.

The Kasi tribe that inhabited Shillong were beautiful people with Mongolian features some had blond hair and blue eyes. They had a matriarchal culture. After getting married, the husband goes to live with his in-laws and his children take on their names. I fell in love from afar, with one blond hair blue-eyed beauty, who was studying in a convent near by. We just smiled at each other when we passed by. I was painfully shy in those days. No guts no glory, but I did avoid living with in-laws.

The Kasi's were happy-go-lucky folks and many would sing and play the guitar in the evenings. They had no written language. One funny thing I noted was that some of them were given names picked at random from an English book. One person I met was called Page 2. Another I was told was named Sunset.

Twenty-six miles away was Cherrapunji, which held the record for the highest rainfall in the world. The maximum rainfall in one year was 904.99 inches and a record one-month of rain measured 366.14 inches. Later, I was in Azzia in Libya, which had recorded the world's hottest temperature, more on that later.

A few of us, decided to hike the 52 mile round trip to Cherrapunji in one day. Surprisingly, it never did rain throughout the hike, lots of damp mist but no precipitation. Memories of exotic water falls, steep river valleys and lush jungles are still in my recall, as well as the tail end of a large snake that lay under me as I sat on a rock wall beside the road. I leaped out of my seat like a shot out of a gun, only to discover that the snake was dead, it had probably been run over and crawled into the rock wall to die. We paid a short visit to a missionary convent in Cherrapunji and were fed by the nuns, who thought we were crazy to attempt to return the same day. We got back to the College after nightfall. It was tricky walking in the dark without flashlights. We were obviously poorly prepared and thoroughly exhausted but exhilarated. In Shillong, there were abundant rumors of a secret society of snake worshipers called the Thlen. One day we witnessed them drumming and chanting around a deep well. We presumed there were snakes in that well. These snake worshipers would lay a curse on a person who would then loose the will to live and die. When examined by doctors, they were found not to have a drop of blood in their bodies. This phenomenon was often mentioned in the local papers and was the subject of a National Geographic article.

A beautiful black horse, with a pronounced limp, due to a swollen fetlock seemed to be abandoned in a field near the college residence. I cured the limp by putting on a bandage soaked in liniment, which was given to me by the College infirmary. Not having a saddle, I rode the horse bareback and named her Velvet. She was high-spirited; especially as I fed her oats, and could turn on a dime, making me suspect that she was once a polo pony. Velvet was a beautiful animal, as good as the horse described in the book, "Black Beauty." On weekends, I would give the kindergarten kids a ride by walking the horse on a halter. The young ones were thrilled and it was a great pleasure to watch their faces beam with enjoyment. I had by then improvised a saddle with a belt and a blanket. Those small kids cried when their parents left them at school for nine months but also cried when it was time to leave the school and go home when winter came. I became their hero, as they really loved to ride on Velvet. Velvet was stabled in a disused kitchen, which the Brothers kindly let me use.

There was one occurrence I will never forget. I was riding on the edge of a soccer field, the same field that a tiger was spotted the year before, when three carthorses charged toward us, and seamed to be attacking.

Little did I know that Velvet was in heat. Front feet of the three horses were flaying around me as they were trying to get mounted. I was sprayed with semen and fell off. Even throwing bricks and stones did not discourage them. Eventually they left, with me in hot pursuit.

I kept a bag of oats in my bedroom. There were eight bedrooms in a row, accessible by several steps leading to a veranda, which surrounded the whole building. All the buildings in Shillong were built above ground on stilts to allow swaying, thus preventing damage due to the constant earthquakes in the region. One early morning, Velvet climbed up the steps, walked along the veranda put her head through my window, trying to reach the oats. Often when I was riding Velvet, a young Red Setter dog would follow us and keep up, no matter how fast we went, thus another stray came into our lives. This one we called Molly, and she soon became the school mascot. The air in the hills was always fresh with a touch of pine scent, in fact Shillong, was known as the Scotland of the East. After coming back from a ride on my favorite trail, I would pat the dog on its back and the horse on its nose. Many years later, I would become so attached to an airplane, that I would pat it on the nose, after it brought me back home safely. That aircraft was the Canberra PR 9. When I was in the Isle of White, to gain working experience with an aircraft company, as part of my college training, I did some poetry writing. This is what I wrote about these two animals.

"A horse, a dog, the open road, is all I crave and ask.
I'll travel far from this abode and sorrows I will mask.
The birds, the trees, the green, green fields, will like a tonic wine,
Restore the drudge the city wields, and now the world is fine.
We stop to rest when ere we will, my horse my dog and I.
To graze to romp to lie so still and watch the world go by.
'Tis a pleasure to lay dreaming of castles in the air
Forgetting all the scheming, that men and women share
My dog comes up and wags his tail, a pat, and he'll be pleased.
A message for all men to hail, when fame they grab and seize.
My horse is uncomplaining, with spirit, I'll not doubt.
But now it starts raining and only I will shout".

Burma was not far away and the Japanese were advancing threatening to invade India. Billeted with us in the University hostel for a short while was a tall handsome coloured American who was with a graves registration unit. His job was to pick up dead bodies from downed aircraft on the border of India and Burma. It was always sobering when he came back from his ventures in the jungle with body bags slung in his jeep.

I was a witness to many strange things throughout my years in India. Men imitating monkeys and naked men, painted as tigers, drawing a large following around them. Whenever two tigers, with their followers met, mock fights ensued and were very realistic. There were plenty of snake charmers to be entertained by. Most were also magicians, who would often pull snakes out of bystander's mouths. I never did see the famous Indian rope trick. Like the Lock Ness Monster, I believe it is a myth. Another strange occurrence took place when we were living in Calcutta. I was visiting a friend, when about 200 rupees, a large sum in those days was missing from his mother's purse, which had been left on her dresser. The police were called in, to no avail. The servants got together and brought in a Sadhu (holy man). He had a large number of servants gathered together in a large circle. The Sadhu placed a shallow earthenware bowl in the center of the circle and filled it with rice. Each of the servants was asked to place both bare feet in the bowl for a few seconds and then step off. When the fifth person stood on the bowl, it twisted sharply and broke. We could see blood on the rice. As soon as this happened, the "culprit" started to run, chased by his fellow servants. They soon caught up with him and gave him a severe beating. Hopefully he really was the culprit.

Life in India was never boring. Growing up in British India was a continuous adventure. The flavor and aroma of the varieties of food do not need to linger in my memory, as I search for and find authentic restaurants, failing which I am told that the curries I manage to cook are to die for.

The first nineteen years of my life could fill a book. The exciting and adventurous life I was fortunate to live, during my early days, was not diminished when, later I started my career as a Fight Test Engineer.

In August 1947 on the day of independence, the family left India on the P&O liner, the SS Strathmore.

This chapter and the chapter of my life in India, is now over.

CHAPTER 4
Learning Curve 1948-1951

We arrived three weeks later at the Tilbury docks at the East end of London. My first impression was quite a shock. I saw row after row of small houses all joined together. They looked like coolie quarters (housing for Indian laborers), with white folk living in them. We were used to living in palatial homes, with many servants at our beck and call. It was quite a culture shock, especially for my mother who, bless her soul, coped very well and even taught herself to cook.

I was registered at the Agricultural College in Chelsea, London. When I got to the College, I noted that it had an Aeronautical section. That's when I switched my intended career path, and do not regret it to this day. With India's independence, my dream of becoming a tea planter had faded away.

The trip to College from North East to South West London was long and arduous. It required arising at 5.30 am and traveling on two buses and three trains. On my very first bus trip I was taken aback by the arrival at the railway station. It was like a stampede to get first in the queue line, which formed at the next interchange. This went on at every interchange. My first impression was that there had been a terrible disaster and people were running away from it. The old and infirm were left far behind and the swiftest got to be first in line. My impression of the docile good behavior of the British citizen was shattered by this experience. Before long, I was in the lead group of those mad dashes. Traveling to Collage a couple of months later was a lot easier when we moved to the new Malden in South West London.

The bulk of the collage students were Norwegian, Dutch, Polish, South African and Canadian ex Air Force personnel, as well as many from the Royal Air Force.

The Aeronautical course encompassed the topics of, aircraft maintenance, structural design and aerodynamics. During the maintenance course, when stripping and reassembling aero engines, I reminded myself that I was once referred to as the "pucka sahib" (perfect gentlemen).

I was finding out what hard work was all about. The dirtier my overalls got the prouder I became.

I joined the University Vandals rugby club, and played almost every weekend, either on the wing or as scrum half. My parents were not too pleased when one day, playing against the London Police, I was knocked unconscious and brought home in an ambulance.

One of my college courses, which lasted a week, was at the National Gas Turbine Establishment in Lutterworth, near Rugby. Frank Whittle, the inventor of the jet engine, later Sir Frank Whittle, ran the establishment. We were given a demonstration of the awesome power of the early Derwent jet engine. In an open field, the circular fuel ring, complete with fuel nozzles was removed from the engine and placed on a stand. Fuel was fed at pressure required for take off power and then ignited. What a spectacular show, a huge flame, hundred's of feet long, sounded like a thousand banshees. Just imagine the current large jet engines with thrust about ten times more than that early jet engine, being subjected to a similar demonstration.

While on the course, a rugby club member came over and asked if anyone could volunteer to play scrum half, for their B team. Both their scrum half's had been injured. I can now always say, that I played rugby in Rugby for the Rugby rugby club.

Some of the college courses were held in Red Hill, which was an R.A.F. fighter base during the war. One of the aircraft hangers housed afuturistic looking, experimental aircraft, called the Planet Satellite. The plane was constructed mainly of a magnesium alloy, to save weight.

The engine was located behind the cockpit and a pusher propeller was positioned back of the tail. On its first flight attempt, the magnesium alloy undercarriage structure, fractured on the take off run, bringing the aircraft to an embarrassing halt. We students were asked to volunteer to put the aircraft together. The engine had been removed and we were to make the prototype cosmetically acceptable for some dignitaries to view. It was a Friday afternoon, and my job was to install firewall baffles in the vacant engine compartment. Each panel had an enormous number of screws. There was a light on a wander lead, in this small, cramped compartment, which gave me adequate visibility. In my enthusiasm to complete the job, I had screwed in about ten screws in the sixth and final baffle and suddenly realized that I was about to entomb myself. Getting myself screwed in was not a smart thing to do. Just as I started unscrewing, the light went out and I was completely in the dark.

Apparently it was quitting time, no one aware of my presence in the aircraft. They all left and turned off the main switch, leaving me in a silly predicament. I am glad that I do not suffer from claustrophobia. Feeling for those screws in pitch darkness was not easy, I managed, half an hour later, to get out of the mess I had created for myself. Lesson number 1, think before you act.

After a thirty-month curriculum was completed, a final six-month period was to be spent, gaining experience at a production or flight operation company. I chose a company called Summerton Airways, who ran a flight training and charter operation on the Isle of White. The advantage there was that they were willing to grant me free flying lessons for the work I would be doing.

One of my first tasks was to conduct a 100-hour maintenance program on a de Havilland Rapide aircraft, an old twin engine biplane.

When it came to greasing the wheel bearings, I was told the grease was in a large tin which I promptly found and completed the job. The grease texture felt unusual and not like any slippery grease I had known, but I slapped it on anyway. The reward for my efforts was to ride on the check flight. It had been raining during part of the flight. Everything was satisfactory and the aircraft taxied back to the hangar. It was then, that we noticed a lot of foam around the wheels. I had used soft soap instead of grease.

Lesson number two, when instinct tells you something is wrong, (the soft soap did not feel like normal grease), double check.

I was in the process of stripping down an Auster aircraft in preparation for its annual Certificate of Airworthiness. Keen to impress, I quickly had the wings off, the propeller off, the undercarriage off and was in the process of removing the fuel tank, which was in the nose of the aircraft, when the owner arrived. He came up and asked me what was I doing to his aircraft. He was here to pick it up. Not for at least a week I said. Just then looking across the hanger, I saw an identical Auster aircraft with the registration number I was told to work on.

Lesson number three, enthusiasm alone, is not good enough.

I was invited to take part in an air show being held at the Navel Air station in Portsmouth. It was a hot day and I was hidden in a tent like structure on wheels, which was towed to the middle of the airfield. Dangling from this structure, held by string, were three empty Champaign bottles. Eventually it became time for my performance. A Tiger moth biplane took off with a crack shot hunter armed with a double barrel shotgun, in the rear seat. As the aircraft approached the target, he leaned out of the open-air cockpit and fired a shot. I awaited the sound of the gunshot, on hearing which, with an almighty, blow, I slugged the first bottle from behind the canvas sheet, using a heavy hammer. The bottle shattered into a multitude of pieces. There was a loud cheer as the gunman supposedly hit his target. On the next run, he fired his shot and apparently missed, as I deliberately did not use my hammer. The crowd groaned. On the third run two shots in quick succession were fired and armed with two hammers, I shattered both bottles' one after the other.

The crowd was awestruck. "Oh what a tangled web we weave when first we practice to deceive." Many years later when, reflecting on this episode, I related the role of the hammer-wielding guy behind the curtain, as that of Flight Test Engineers, smashing away problems, hidden behind a curtain. The hotshot hunter's role was that of the Test Pilots, always in the limelight. A limelight much deserved by some and not by others.

My father, who had retired from his Chief Account post at the Indian railways just before we left India, was now back at work as Financial Adviser to the new formed government of Pakistan after being courted by previous colleagues to assist the Pakistan Embassy in London.

My father's entry into the diplomatic core, meant that my parents attended many cocktail functions, including garden parties at Buckingham Palace.

While I was at the Isle of Wight, my father paid me a visit and I took him for a flight in an Auster 5 light aircraft. He had never flown before and it was a thrill for both of us. He was always proud of me, and I was so glad to see his excitement. He even got to handle the controls for a bit. He passed away a couple of years later.

Studying a little, playing a lot of rugby and attending dances at as many teacher and nurses training colleges I could find, I completed the three years it took, to obtain my Diploma of the College of Aeronautical Engineering. Armed with this credential, I imagined, only the skies would be the limit, applied for my first job and started my career in aviation.

CHAPTER 5
SAFETY TRAINING

A pre requisite to test flying on several high performance military aircraft, is that medical examinations and safety courses had to be undertaken.

SEA SURVIVAL COURSE

This course was conducted at a search and rescue unit located in Plymouth in the south of England.

All crew flying in certain types of military aircraft, carry a small dinghy pack attached to the base of a parachute harness. The pack is located in an indent on the crew bucket seat and acts as a seat cushion. If bailing out over water, quick release clips are released, allowing the dingy pack to detach and dangle down, held by a lanyard. On contact with the water, the dinghy can be inflated by pulling a tag attached to an air pressure bottle. If any inadvertent inflation occurs aboard the aircraft, a dinghy knife is supplied to each crewmember, enabling him to puncture the inflated dinghy. An inflated dinghy in a confined space can be the cause a serious emergency.

The dinghy is made of a flexible rubber material and is coloured a bright orange. When inflated it will support only one person, and contains an emergency kit. The kit contains a search and rescue homing beacon, a whistle, an emergency light, a marker dye, a canvas bailing cup, shark repellant, a fishing line and a first aid kit.

Initially dinghy drills were held in the comfort of swimming pools. Dressed in swimming trunks, we would be acquainted with the contents of the dinghy, and then jump in the water holding on to the dinghy pack. We would inflate the pack and be shown how to enter the dinghy. That was all there was to it, simply a fun experience and an opportunity to go for a swim.

A near fatal incident had occurred, when an English Electric Test pilot ejected from a P1 Fighter aircraft. After touching down in the English Channel, it was suggested that he was either unaware of the ability and use of safety equipment supplied, or the shock of bailing out, using his ejector seat and also the frigid cold and angry sea, caused him to loose his survival instincts. As a result he was lost at sea for three days and was lucky to be found alive. This episode resulted in a more realistic sea survival course, which I was to attend. The courses in Plymouth were held all year round. It became my turn to attend the course, in the middle of winter.

We were required to wear all the safety equipment used on the aircraft we were flying in. In my case, as a crewmember on the Delta wing Vulcan Bomber aircraft, I wore a parachute harness, dingy pack, May West, (Personnel flotation device), a flying suit, a pressure waistcoat, G suit, and a pressure helmet. This was a far cry from the swim trunks worn on my last dinghy course in the comfort of a warm swimming pool. The day before my christening in the English Channel, we were given safety lectures, stressing the importance of the equipment contained in the dinghy and especially the emergency locator beacon located in the May West. A Surprising amount of emphasis was given to the importance of maintaining a strong will to live. This I thought was unnecessary, believing that survival was a natural instinct.

On a cold early January morning, a high-speed rescue boat took us far out to sea. There were 15 of us on that boat; I was the second person to jump in. We were told that, in approximately two hours, some rescue helicopters would arrive, lower baskets and hoist us up. I thought that would be exciting. It was to be a training exercise for the helicopters as well.

After leaping out from the back of the boat which was moving at a fair speed, I managed to locate the lanyard and pull the pack towards me, inflate the dinghy and struggled to climb in. Cold and somewhat exhausted, I was bounced around like a cork. The English Channel is seldom calm. The more the winds increased, the more I was bounced around. In all the flights that I have undertaken throughout my career I never did require the use of sick bags that were carried on all test flights. But, boy did I get sick in that dinghy. I heaved constantly and soon was wallowing in a toxic and disgusting looking mixture of ice-cold seawater and vomit. I thought I was bringing up my insides and only days later, realized I had a meal the night before that included dark red beetroot.

The winds increased, the waves grew bigger and bigger. At the start of the exercise, I could see the odd dinghy, as both would be at the crest of a wave at the same time. Soon no one appeared in sight as we had drifted far apart, even though, as the waves grew bigger I was up on a higher point. Thinking it was time to be rescued, I attempted to operate the search and rescue beacon attached to the May West. My hands and fingers were so cold that I could not even open the pouch that held the rescue beacon. A storm had increased in intensity, the increasing wind and waves, unknown to me, prevented the rescue aircraft, leaving base. I kept attempting to throw up but to no avail, there was nothing left in me. Water kept splashing into the dingy, which I attempted to bale out but eventually quit, tired and exhausted. I sat in that puddle of ice-cold water and vomit, head bowed, and hate to admit it, but I just wanted to die. In vane I searched the sky for my rescuer. The storm raged on all day and was growing in intensity. Violent shivering and chattering teeth, indicated that hypothermia was setting in. All I could do was to rub my hands together and rock forward and back. I prayed that the gods would not abandon me. It was getting very unpleasant, the sun had set below the horizon, the fast approaching dark outside, matched my inner darkness when I heard the distant horn and after what seemed like an eternity, the rescue boat loomed up, bathing me with its searchlight as I bobbed up and down. The shining light brightened my inner despair and soon I was hauled aboard. Wrapped in a blanket, I was given a cup of hot coco, which immediately left me and joined the rest of my stomach left in that cold and angry sea. The second person to jump in, I was the last of the fifteen to be plucked out and rescued on what had become a real rescue mission. I was informed that the weather had become unsuitable for a helicopter rescue and soon would also be unsuitable for the rescue launch, due to the increasing storm and impending darkness. Thank goodness for my rescue, I could not imagine spending the night and possibly several others in that dinghy. Once again, the Gods were with me.

After this, I surmised that advice given to maintain a will to live, is easier said than done. I also will admit that this was a truly realistic training exercise, never to be forgotten.

HIGH ALTITUDE INDOCTRINATION

Having been chosen as flight crew for the Vulcan bomber aircraft and later, the high altitude Canberra P.R. 9, Photo Reconnaissance aircraft, it was necessary to be cleared for high altitude flying. First, a stringent medical examination must be passed. This was followed by an evaluation in a pressure chamber operated by the Institute of Aviation Medicine located in Farnborough in the U.K.

The pressure chamber was a small cylinder like structure with a heavy round submarine type entrance hatch with dark rubber seals. It was approximately ten feet long and six feet wide. When the entrance was clanged shut, you felt that you were in a tiny submarine. There were four of us in the chamber. One was an Air Force Navigator who, soon after the hatch was closed, became very agitated and showed distinct signs of claustrophobic behavior. He was quickly removed. When the hatch was closed again, the height in the chamber was reduced to fifteen thousand feet. Here we were given a demonstration of anoxia. We were asked to remove our oxygen masks and start writing numbers backwards from sixty. As soon as safety personnel watching from the outside through small windows, and noticed our lack of concentration and inability to continue writing down numbers in descending order, we were told to re don our masks and return to oxygen breathing. We were brought down to sea level pressure then rapidly up to forty thousand feet. We were wearing pressure breathing oxygen masks. Under normal breathing, a person consciously breathes in and unconsciously breathes out. Pressure breathing requires an unnatural breathing process, i.e., unconscious breathing in and conscious and very deliberate breathing out. Pressure breathing is essential at altitudes above thirty eight thousand feet. Oxygen regulators control breathing by increasing pressure with an increase in altitude. Pressure breathing will no longer be possible at heights above fifty thousand feet. This is due to the fact that the high oxygen pressure required at this height will make it impossible to exhale and thus would cause suffocation.

The chamber altitude was maintained at forty-thousand feet for a sufficient long period to note how susceptible we were to decompression sickness, often referred to as the bends. Bends are caused by a rapid change of air pressure and can occur anywhere above fifteen thousand feet. Deep-sea divers often encounter them if they ascend to the surface too quickly. This pressure change can create gas bubbles in the blood stream and tissues and can cause severe pain in joints often resulting in many serious consequences.

Two of us showed symptoms of the bends, near the end of the session.

One had severe shoulder pains, and I had severe pain in one knee.

Soon after the chamber pressure was increased to descend to sea level, pain immediately disappeared and the experience was over. In order to reduce the likelihood of getting the bends, we were advised to breathe pure oxygen for at least half an hour prior to conducting flights to altitude.

This was required to purge nitrogen from the blood stream. Not too many of us paid attention to this recommendation assuming that we would be breathing oxygen immediately after engine start up and during pre flight checks and while taxiing to the runway.

I was now qualified as aircrew for the AVRO Vulcan delta wing bomber, which would not be flying above fifty thousand feet.

Later, I paid several other visits to the pressure chamber to qualify for the extreme high altitude required for operation in the Canberra P.P. 9. Flight-testing on this aircraft would be undertaken at heights above sixty-five thousand feet. If pressure were not applied to the body, death would occur in a few seconds. Blood and body fluids at such low pressures would boil, resulting in instant death also pressure breathing would not be possible. Hence the pressure suits.

In the early days of my high altitude flying, pressure suits were not fully developed. Obtaining satisfactory joint movement with the suits pressurized, was a problem, which would be solved later. Partial pressure suits were developed and were available in 1958 when the high altitude test program on the Canberra P.R.9 commenced. The Royal Aircraft Establishment developed my first partial pressure suit. It comprised of the following components.

A g-suit, which pressurized the lower abdomen and legs but not below the ankle. A sleeveless, pressure jacket and a pressure helmet, with a fixed visor. The helmet contained a pressure breathing oxygen mask and a sick port, (for obvious reasons). The neck, arms, ankle and foot were not pressurized.

Custom fitted with the partial pressure suit, I returned to Farnborough to be indoctrinated with my new pressure suit. Alex Roberts, who was selected to be the pilot I was to partner on the PR 9 test program accompanied me.

Before entering the pressure chamber, sitting on a chair, we were hooked up to a pressure supply. The doctors conducting the course had already been informed of the estimated length of time, required for the aircraft to conduct a rapid emergency descend from a maximum altitude, down to 48,000 ft, where suit inflation was not required. This time was estimated to be four minutes. The suit was inflated to the pressure required for four minutes. At first, it was just uncomfortable, and then it was increasingly painful. Four minutes was an eternity. The pain in my arms, which were not pressurized, was excruciating. When four minutes were completed, the suit was deflated and pain immediately ceased.

The next step was for both of us to enter the chamber, with a doctor similarly equipped and were then taken up to a height of twenty-thousand feet and subjected to a sudden decompression to sixty-thousand feet. As soon as the chamber pressure was suddenly reduced, the chamber fogged up, the pressure suits inflated and the indoctrination was over.

After exiting from the chamber, we heard a group of doctors, discussing an incident from an indoctrination conducted the previous week. They were whispering quite softly but the words, autopsy and embolism were heard. I believe someone may have died on the previous session.

The following day, I noticed my arms and part of my neck had turned black with a tinge of yellow, as though I had suffered severe bruising. I understood the reason for this was that the pressure created by the suit, caused the blood in arteries in the non-pressurized areas to ooze out of the artery walls, thus causing the bruising.

Later, a sleeved pressure jacket, developed to allow adequate arm movement, became available. Also, a new version of a pressure helmet fitted with a movable visor, manufactured by Baxter, Tailor and Woodhouse, was available. This visor when unlatched by a rapid change in air pressure, snapped into place sealing the open face area. On occasions when exhaling heavily, the visor would snap shut like a rat trap, any fingers in the way would get a painful rap.

A return visit was made to the pressure chamber to check out this new equipment.

There were several Pilots and crew members on this course. The new helmets had a zipper at the back, which allowed quick removal simply by tugging on a strap located behind. The face visor was designed to close shut in the event of a sudden decompression.

As with the previous occasion, pressure was applied to the suits outside the chamber. This was done one person at a time. Because of the introduction of full arm sleeves, the allotted time with the suit inflated, caused significantly less pain than that encountered previously with the sleeveless jackets.

When a Handley Page Company test pilot was first subjected to suit inflation, his helmet commenced to rise up from his head. He was in much pain and discomfort so pressure was reduced immediately. The reason this happened is interesting. Before being supplied with an individual helmet, careful measurements are taken. One of these measurements is neck circumference. It turns out that each helmet is built with an area on top of the head left un-pressurized. This area is equivalent to the neck area of the owner of the helmet dimensions, resulting in this incident.Picture a balloon being inflated and placed on a greasy finger. It will always try to blow off the finger and fly around until deflated.

This unfortunate pilot, who was rather a large man, was given a helmet with the wrong dimensions.

It came time for decompression in the chamber. As usual the doctor was in there to look after our welfare. The moment of decompression, when the chamber pressure was rapidly decreased to sixty-thousand feet, in the ensuing fog, all our suites and helmets became pressurized except that the doctor's helmet came off his head and showing no panic he attempted to hold it to his head. This was a real emergency and the outside chamber safety crew reacted rapidly to reduce altitude in the chamber. It was hard to believe, but that doctor showed no signs of any serious damage done to him. What had happened was that the locking tab on his zipper at the back of his helmet was not in the locked position.

The quick release zipper released, un-commanded, as soon as suit and helmet were pressurized. I never did know if he suffered any after affects from this incident.

The conclusion reached from the Indoctrination experience is that the early pioneering venture into extreme high altitude flight, was indeed hazardous.

CHAPTER 6
AV Roe Woodford Cheshire U.K. (1951–1953)

My first job application landed me a job with the A.V Roe Aircraft Company.

In 1951, I was employed as a Junior Flight Test Engineer with A.V.Roe in Woodford, Cheshire. This was the start of a 48-year career in flight-testing, military, civil and research aircraft.

Alliot Verdon Roe started an aircraft manufacturing in Manchester in 1910. Aircraft sections built in Manchester were transported by road to an airfield in Woodford in Cheshire. Before presenting the chapter on test experience with AV Roe a job description of the Flight Test Engineer job description is in order.

Flight-testing is a dangerous process, since the reason for testing is to demonstrate the ability to fly safely in extreme conditions. The duty of a Flight Test Engineer is to plan, direct and participate in test, always keeping safety in mind. He records valuable data and report the results of flight tests. The efficient use of flight time grows with experience.

He is required to have knowledge of the test aircraft systems and flight limitations. Safety is paramount and must be continuously be in focus. The luxury of referring to operating manuals is not available as these are yet to be written. He is also required to have clear understanding of Regulations. The crew positions he would occupy, varied with the aircraft being tested. This ranged from a co-pilot seat, jump seat, (seat between pilot and co-pilot), navigators station or an instrumentation station designed specifically for his tasks. On some tests, especially when tests are to be conducted on single seat aircraft, he is required to direct the tests by radio contact from the ground.

The test engineer is required to record test data on instrumentation appropriate to the test and maintain a log during the tests. This log would later allow the matching of test points with the recorded instrumentation. Test configuration, test speed, test weight, test center of gravity and pilot's comments were also to be logged.

On most prototype aircraft, a development and certification test program, on average takes two years and 1500 flight hours to complete.

There can be as many as five aircraft used on a test program. Some test aircraft, may have as many as 2000 items, instrumented and recorded.

The recording and play back of test instrumentation is an essential ingredient for all test flying. As my career marched on, I saw an amazing growth in instrumentation technology, from tape measures, spring push/pull force scales and photo recorded instrument panels to computer generated data acquisition systems.

Tests include, handling qualities, engine, electric systems, hydraulic systems, fuel systems air conditioning and aircraft performance also tests to clear operation under hot, cold, high, severe icing conditions as well as wet and slush covered runways. The final certification task is a function and reliability test program, which is required to demonstrate safe operation of the aircraft and all systems during 200 hours of flying, on typical operator routes. Functions must also include demonstrations of emergency procedures.

Military Bomber and Fighter aircraft are tested to military standards. In some areas of operation, less stringent standards apply, as compared with civil requirements.

Research and accident investigations are conducted to Company standards, geared to enhancing knowledge and improving safety.

AVRO SHACKLETON (1951)

The AV Roe Shackleton, was an R.A.F. Costal Command aircraft, capable of many roles such as a submarine hunter, mine laying, search and rescue (capable of dropping very large inflatable rafts), and photographic survey. Four Rolls Royce Griffin engines powered the aircraft, each with two contra rotating propellers. The aircraft had an extremely long range and was capable of staying aloft many hours. On long missions, an additional crew was carried to operate a second shift. As a demonstration of a long submarine hunting mission capability, a twenty-six hour non stop flight was demonstrated.

My first flight test experience was in April 1951 in the Mark 1 Shackleton. It was not a good experience. The noise and vibration were extremely high, and I could not understand a single word spoken on the intercom, until a crewmember adjusted the squelch on my transmitter/receiver. This adjustment fine-tuned the frequency, thus enabling me to gain faint recognition of the odd word here and there. This was a familiarization flight for me. The task in hand was to investigate "Stick Force per g" (an investigation to establish the pilot's effort to pull up the nose of the aircraft and subject it to an increase in g force. The pull force that the pilot exerted must be shown to increase, acceptably with increasing airspeed and g. In simpler terms, the rate of nose up rotation must be proportional to the force the pilot applies and this force must increase with speed. If this did not occur, it would be very difficult to control the aircraft.

It was a roller coaster ride and I wondered if I could ever get comfortable in that noisy, vibrating and sick provoking environment. The noise and vibration in the Shackleton has been described by operating crew as the result of a hundred thousand rivets, screaming to get free and rip the aircraft to pieces.

Tests were conducted at many speeds. At each test speed, the nose would be pushed down, increasing the speed, then pulled up rapidly.

Once the test content was understood and crew voices on the intercom became more understandable, I lost the early feeling that I would never make it in this job.

As I started my career, I felt very humble, thinking I would be surrounded by a number of extremely smart people (boffins), whose standard of excellence, I could never reach. As time marched on, I came to realize, that really smart people were few and far between. A small handful of personnel were really the key to a company's success. One such key person at Avro's was the Chief Flight Test Engineer, Stan Nichol, a person who taught me a lot about testing. The rest of us were average, improving with experience and succeeding only by maintaining our curiosity and always keeping safety in mind and recognizing the importance of communication skills.

One of my early tasks was a simple one. A runway expansion was to take place and a photo grid survey of the airfield and surrounding company property was to be undertaken for the benefit of the building contractors. All I had to do was to press a camera button when someone called out "now". It took about an hour to complete a grid survey. My task was done, until I received a call from the photographic department.

All the photos were blank. When someone asked me if I had opened the camera doors, my heart sank and I knew I blue it. The lesson here was that time spent in preparation for tests and gaining an intimate knowledge of the aircraft systems was essential. Do not expect to be spoon-fed.

Development flying was intertwined with production aircraft tests. Each production aircraft undergoes a checkout test program. Tests include checking the speed envelope from stalls to the maximum dive speed. Cruise speed and climb performance is recorded. All systems are checked including engine starts in flight and emergency functions.

A production aircraft takes on average, four flights before it can be cleared for delivery.

The Mark 1 Shackleton had been fitted with a chin-mounted radar, this was to be replaced with a retractable belly mounted radar. To clear this change in advance, the prototype aircraft was modified to remove the existing radar and install a retractable mockup radome in the belly.

When lowered in flight, the radar protruded a good five feet below the bottom of the fuselage. It was a mockup, oval in shape, the outer skin was made of plywood and located mid fuselage. With the radar unit not installed it could easily accommodate a man standing with his arms stretched. I was to be this man.

With the random fully extended, severe buffeting and vibration occurred, which could not be tolerated. Measurements were required. I was lowered into the extended radome and armed with a handheld, spring wound vibration meter an instrument not unlike a seismograph. As airspeed was increased, the buffeting encountered can only be described as that felt by an ice cube in a cocktail shaker and the noise level as being inside a bass drum, with the drummer gone berserk.

I could not believe that the plywood outer skin of my "potential coffin," would remain intact, but it did. I managed to get some frequency and amplitude readings before getting out of that hellhole.

A lot of my three years spent at AV Roe were occupied with production check out flights on Shackletons Mark 1,2 and 3. I got to know that noisy aircraft very well. I especially enjoyed, getting into the rear gun turret on the way back from flights and swinging my seat up, down and around, enjoying the view and taking aim at imaginary targets.

A heavy landing test program, created an interesting demand for touch down sink rate measurements. We happened to be at the site and with the company that built the Lancaster Bomber and installed modifications to the 19 aircraft that took part in the attacks on German dams. The rotating bomb technique, pioneered by Barnes Wallis, relied on maintaining an accurate 60 ft height over water. Operation Chastise was the official name for the attacks on the German held dams. From a total of one hundred crewmembers, fifty-three were killed and three parachuted out and were made prisoners of war. The dam busting raids caused devastating floods, even more devastating than hurricane Katrina and slowed down the German war effort. The attacks on the dams were made famous by the film "The Dam Busters". Two searchlights were used during the raid to ensure accurate height keeping, over the water.

When the searchlight reflections over water, melded into one, the height was exactly 60 ft.

A single searchlight was located, in the bomb bay of the Shackleton and a camera with an accurate time base was positioned to measure the diameter of the light image, reflecting on the runway. Time and diameter would give an accurate sink rate. It became necessary to calibrate the searchlight image as it changed with height above ground. The instrumentation department was given this task. I happened to walk by a group who were armed with shovels. They were marking a circle, in preparation for the calibration. The aircraft was required to wheel across the circle and then out again. Each time they would measure the reflected light diameter, then dig a foot or two deeper and keep repeating the process until the hole I presumed would be wider than the wheel span of the aircraft.

I casually remarked that it would be a lot simpler and less labor intensive to remove the searchlight from the bomb bay, place it on a trestle, provide a power supply and shine it on the hangar doors, moving back a foot at a time. This they did, rather than a lot of digging.

I use this example to demonstrate that often when caught up with details, we often miss the simple approach to solve problems.

Occasionally, I was asked to take part in tests carried out in the Avro Ashton aircraft. It was one of the first four engine jet aircraft to be built. Only six were manufactured and all took part in research programs ranging from bomb aiming and dropping to being used as flying test beds for jet engine development.

We were checking out a new wing pod installation, which was to be used for carrying dummy bombs when suddenly, a fuselage window blew out causing a loud bang accompanied by a fog. All aboard were carrying portable oxygen bottles and used them during a rapid descent to a safe altitude.

The bomb aiming position in the Ashton had a very large window approximately 4 ft x 3 ft, located forward, in the belly of the fuselage.

There was a concern with thermal stresses on this window. My task was to measure window surface temperatures at various locations on the window. Armed with a sensitive temperature-measuring instrument and wearing a parachute and safety harness, in case the window cracked.

I was lowered over the window. It was an unusual feeling looking down with nothing obscuring my view. It was a genuine bird's eye view.

There were several other projects taking place in the Flight Test Department. The Athena, an advanced trainer with a Rolls Royce Merlin engine, the one powering the famous Spitfire fighter, used during the battle of Britain and a four-engine jet bomber called the Sperrin.

The two-man crew of the Athena were often subjected to forced landings. Two landed on golf courses and another in a farmer's field. Oil temperatures above limits caused engine shutdowns and subsequent emergency landings. No one was ever hurt and emergency phone calls announcing the crew whereabouts was a common occurrence.

Single seat delta wing research aircraft flight tests were also taking place; my only contribution was taking part in data analysis.

When I was told that I was selected as a crewmember for the new Vulcan delta wing bomber, I was overjoyed.

Adventure in the Air — 51

Avro Shackleton
Bomb Doors Open

Avro Vulcan Bomber

CHAPTER 7
AVRO VULCAN (1952)

The British Government issued a contract to supply heavy bombers to the R.A.F., capable of carrying the atom bomb and capable of flying at fifty-thousand feet at a speed near the speed of sound.

Three companies were contracted to develop such a bomber aircraft.

The Vickers Aircraft Company the Handley Page Aircraft Company and the A.V. Roe Aircraft Company.

This created the V-Bomber trio, the Vickers Valiant, the Handley Page Victor and the AV Roe Vulcan. As a back up, the Short brothers and Harland were also contracted to build an aircraft called the Shorts Sperrin.

This I believe was an amazing display of the panic caused by the cold war fear of a Russian attack and showed the determination by the British Government to gain a superior atomic deterrent capability. With no regard to the immense costs, the desire to hold the lead in the arms race, led to a double belt and double brace approach. Failure was not an option.

The Valiant, using existing technology was the first to fly in1951.The Victor and the Vulcan, risking new technology, would fly later.

The Sperrin was to be available as a stopgap in case the Victor and Vulcan, with risky new wing designs, failed to succeed.

The race was on between the Victor and The Vulcan not just to be next to fly, but to make an appearance at the 1952 Farnborough air show. Both aircraft were close to their first flight.

The hold up on the Vulcan was the complicated fuel system. There were many fuel tanks installed in that large delta wing. A computer system was devised to control fuel in order to ensure that the center of gravity of the aircraft would remain within limits. In order to make the air show on time, the fuel system was disengaged and capped off. Large fuel drums were installed in the bomb bay and plumbed to supply the engines with fuel.

Engine runs to check all systems, were delayed until the pilots and co-pilots Martin Baker ejector seats were installed. When the seats eventually arrived and were installed, it became evident that there was insufficient clearance to allow the cockpit canopy to be closed. Interference with the canopy was on the co-pilots side. The solution was to remove the co-pilots seat. The intended Rolls Royce Olympus engines were not available in time. Rolls Royce Avon engines, with less thrust and which were readily available were hurriedly installed. On August 1952, without test instrumentation, the first flight was conducted with the pilot as the only person aboard. After take off, both undercarriage doors fell off. A few days later the Vulcan successfully demonstrated at the Farnborough Air Show with a lash up fuel system, less powerful engines and only one person aboard in a cockpit, designed for a two man crew. The media hailed this large aircraft flown by only one man, as a wonderful accomplishment. Little did they know there was no seat for the copilot.

This display of desperation to be the first at the air show was a sign of the times. The perceived imminent threat of world war 3 and the race to be first out of the gate was fueled by intense competition and the promise of a fat contract from the Ministry of Defense. The Company was walking a fine line between safety and glory. However all's well that ends well.

Note: The Victor did not appear at the show until the following year.

After the air show, the aircraft fuel system was completely installed and the fuel, temporally installed in drums in the bomb bay for the air show was replaced with water in those drums and used as ballast for shifting the center of gravity to accommodate test requirements.

The Co-pilots ejector seat was installed as well as a full test instrumentation package. This package contained several oscillographs, which traced a time history of fast moving parameters and a very large photo panel, which contained about seven hundred instruments. Three cameras were used to cover the large instrumented area.

An army of women, approximately thirty in number, conducted data reduction. The girls spent each working day in the dark, in front projected images of instruments, recording time and prescribed instrument values. It was a very boring job, yet they worked tirelessly. Why only women were given the task of film reading, I do not know. In those days, females, who worked in an engineering field were a rare breed; perhaps it was a start to overcome the gender gap.

Soon after the test program commenced, the program required operation at high altitude. The pressurization system was not yet available and it became evident that the new oxygen regulators supplied for the program were unreliable and occasionally failed to provide adequate oxygen. This hazard had to be dealt with. Improved regulators would not be available for quite some time. To continue the program we had to rely on the buddy system. The co-pilot would watch out for the pilot and vice versa. A second test engineer would, always accompany the test engineer, located below and out of sight from the pilot crew up front. Both would watch out for each other.

One of the tests we conducted, was a fuel system test, which required a rapid climb to high altitude to record fuel tank pressures and check the adequacy of the tank vent system. As we reached an altitude more than forty thousand feet, I noticed that the test engineer, Stan Nichol, sitting across from me was struggling to remove his oxygen mask. He did not respond to my call to stop this apparent foolish action, I reached over to him forcibly shoving his mask to his face while he fought me off. When I noticed slime on my hands, I realized that he had gotten sick in his mask and unwittingly I was trying to gag him. As soon as the emergency was declared the aircraft made a rapid descent and returned to base where an ambulance lay in waiting and collected Stan for a check up. He was fine. It goes to show that the problems you encounter are not always the fault of the aircraft. Humans often contribute to problems.

Miscellaneous tests on a Lincoln Bomber, a post war version of the Lancaster and some directional control tests on an Anson, twin engine trainer, which had been modified for the Indian Air Force, rounded out my work at AV Roe.

I was about to get married to Brenda, a Lancashire girl and heard that an Aircraft Company in Northern Ireland was offering housing to suitable applicants. Good, affordable housing, in those days, was hard to come by, so we grabbed the chance to start our new life together, across the Irish Sea.

With some regrets, I left AV Roe and started life in Northern Ireland.

CHAPTER 8
Short Brothers and Harland, Belfast, N Ireland. (1953- 1956)

My application to Short Bros and Harland was successful and after a honeymoon in Bray near Dublin in the South of Ireland, we headed north to Belfast, where I joined the Company and was given the keys to a detached house on the outskirts of the city. The house was completely empty and devoid of any furniture. We had spent the last of our savings on the honeymoon and could only afford a small rug and a radio. We pasted brown paper over the front room window, for privacy and slept on the rug under which we placed woolen clothing to provide some softness. The "bed" was soft in a few places and lumpy in others, affording a poor night's sleep. Week by week we added furniture and after a few months, we were quite comfortable.

Shorts Flight Test Department, at the time, was located in a small hut located on the airfield away from the main plant. The pilots were located nearby in a corner of a hanger. A total of ten personnel occupied the Flight Test Hut. There were six test Engineers, three Instrument laboratory Technicians and a Secretary. I managed to get a few flights on the Sealand, an amphibious twin-engine seaplane. It was my first experience with water handling tests. I witnessed a porpoise so severe that the water spray over the windscreen turned a solid green colour, indicating that the nose of the aircraft was below the water surface. The term porpoise is an apt description of bobbing up and down on the surface of water just like the swimming action of a porpoise. The sensation reminded me of a camel ride many years ago. The back and fourth motion was more violent than sensed on the camel.

Miscellaneous pre delivery tests were conducted on a Shorts Sturgen, which was a twin engine target towing aircraft used for air to air gunnery practice, and a Sunderland flying boat which was modified and was required to be checked out prior to delivery to South Africa. It was an aircraft similar to the ones I saw many years previously conducting a high-speed water operation on the Brahmaputra River in India.

SHORT SEAMEW

I was assigned to the Seamew test program and teamed up with Wally Runcenman who was to pilot the aircraft. He hailed from New Zealand.

The Short Seamew, was a single turboprop aircraft designed to operate from an aircraft carrier and destined to carry torpedoes and hunt submarines. This aircraft gave me a good learning experience and taught me many lessons.

On it's first flight, the aircraft landed short of the runway on rocky ground and suffered significant damage. In retrospect, I realize that this incident could have been avoidable, if pre first flight high-speed taxi tests were conducted. Taxi tests would have revealed that when throttling back the engine to idle, the thrust at idle was too high and should have been adjusted before first flight. When the aircraft came in to land, the pilot could not achieve adequate speed and sink rate to allow a landing, so prior to the first landing, the one and only engine had to be shut down, leaving no capability of going around if required. As a result, the aircraft undershot the landing and touched down in a rocky area, causing damage to the undercarriage.

It soon became evident that the roll response was inadequate and that the pilot's workload to pick up the wing was too high. The stringent requirements for an aircraft to land on the deck of an aircraft carrier were not met. This deficiency led to an intense lateral control improvement test program. Being the only test engineer that did not require a sick bag during the large number of rolls conducted on each flight, I remained the only one assigned to the program. Six months later, having conducted approximately one thousand rolls, the final modification allowed the aircraft to just meet requirements.

Now longitudinal stability tests began and problems needed fixing. This took another four months to fix. Next comes a directional stability program and you guessed it, problems needed fixing, requiring major modifications. This took another four months.

After all stability and control and remaining test programs were completed, a check with the bomb doors open was conducted. This was the first time the doors had been opened. Everything went fine until I noticed my feet were getting hotter and hotter. My seat position was behind the pilot. My feet were on the floor; the pilot's feet were on the rudder pedals. Realizing that the floor was getting hot, I suggested that we closed the bomb doors. The floor remained hot, so I put my feet up on the bulkhead in front of me. I still felt the heat and realized that the crepe soles on my shoes had absorbed a lot of heat. I took my shoes off and soon had happy feet. The problem was obvious as the engine exhaust was just in front of the bomb doors, yet not one of us anticipated the problem. When the bomb doors were open, hot exhaust air would circulate in the bomb bay causing the floor panels to heat up. The pilot, being ahead of me with his feet on the pedals did not get the hot foot. A major modification would be required to fix the problem.

The lessons I learned later from this program were, never to concentrate on one subject with the exclusion of others. The early flight test program must examine all subjects as early as possible. At least more than one test aircraft should be used, each aircraft conducting different tasks.

After development tests were completed, dummy deck landings were conducted and the Royal Navy accepted the aircraft. Seamew production aircraft were each checked out, flown to Scotland and then moth balled and stored for potential use, and were never used. It was a lot of effort for naught.

CANBERRA

Short Bros And Harland in Belfast Northern Ireland was now building several versions of the English Electric Canberra Bomber under license.

I participated in many production test flights on the B-2, B-6 and B-8 versions of this rugged twin jet bomber, which was the mainstay of RAF Bomber Command.

Production testing got to be quite a routine. Each aircraft went through the same exercise i.e. check function of all systems and equipment, check for buffet and control flutter at maximum speed and Mach number and check handling qualities including stalls. Many repeat flights were required to check out fixes to problems encountered.

The Canberra aircraft production test program was conducted at Aldergrove, which was at that time was a military base. Aldergrove was near Belfast and is now the Belfast civil airport.

There were four production test pilots and four flight test engineers assigned to production aircraft clearance check flights. I had my own flight equipment and was directed to a flight safety room where I picked up a parachute and a bright yellow life jacket (May West). My first Canberra flight was exhilarating. I was pleasantly surprised at the low noise and vibration levels. It was a busy flight as we went through a long checklist covering system and handling tests. When we got around to conducting the dive to check the maximum speed limit, which was around Mach .84 (645mph), I got my first experience of heavy buffet.

If it were not for the experienced pilot advising me that the buffet was quite normal, I would have thought that the aircraft was about to come apart. The Canberra is a tough old bird. While back on the ground and still sitting in the aircraft, writing snag (defect) sheets and marking tests that were outstanding and tests to be repeated, a Canberra taxied and came to a halt beside me. Out of this aircraft appeared an irate pilot, whose face was red and distorted with anger. With the words " you took my f—-ng May West you f——ng B—— rd", he slammed his Bone Dome (hard helmet) to the ground, it bounced about ten feet up in the air and eventually came to rest. It may have been a superstition he had about not wearing his own precious May West, but there was no call for such behavior. At that time, he was a production test pilot. Later on he was made the Director of Flight Operations and would become my boss.

Every time I came across that man, the words "Prima Donna" came to mind. The episode had to be forgotten as we were thrown together often on production aircraft test flights.

Many a time, he threatened to fire people for little or no reason. After he had been promoted, he chaired many post flight debriefings.

On one occasion, a production aircraft had conducted a lot more than the usual number of flights and still was not cleared for delivery. A friend of mine Stuart Nicholson was unfortunately the test engineer allocated to that rogue aircraft. I have known Stuart for more than half a century. He was a fine test engineer and followed me to Canada together with Jock Aitken who was the first to go to Canada. The three of us were referred to when there as the Irish Brigade. At the debriefing, Stuart contradicted the Chairman over the legitimacy of a snag. Stuart was right, the Chairman was wrong. Contradicting a Prima Donna in front of his subordinates was a recipe for violent reaction, and Stuart was fired. I got to hear of the incident, pleaded Stuart's case and had him reinstated.

The completion of the Seamew test program coincided with the completion of all other test programs, resulting in complete and sudden lack of activity. The de Havilland Comet four engine jet passenger aircraft was being manufactured under a license agreement and was just about ready to make its first flight, when several spectacular crashes caused the program to be abandoned. Accident investigation revealed that the crashes were due to metal fatigue of the fuselage structure, caused by cabin pressurization cycles.

The Comet pioneered passenger jet flying and the knowledge obtained from the accidents has resulted in elimination of metal fatigue that caused the crashes.

With nothing to do we found ways of keeping ourselves occupied. Someone came up with the idea of making insect powered model aircraft, with matchsticks, paper and glue. Our power source consisted of flies, dragonflies and bluebottles, with their legs stuck on using office glue. We refrained from using bees and wasps after suffering a few stings. Soon we had single engine and twin-engine tiny models flying all over the office. Take off time was unpredictable and depended on the whim of the power source. Lack of an acceptable thrust to weight ratio was evident when some models would stall and others could only muster a powered descent. Twin engine flying was often erratic as the port insect did not apply the same lift and thrust as the starboard Insect. Points were given for the longest flight time, the furthest flown and the best aerobatic display. Upon reflection, it was no more cruel than placing a worm, minnow or frog on a hook to be devoured by a hungry fish.

No one objected when lunch took several hours. Often we went to the center of Belfast, which was only couple of miles away and had a pie and a pint or a Ploughmen's lunch with a foaming glass of Guinness at Moonies pub.

In spite of the lack of activity and no signs of new projects on the horizon, there never was a fear of lay offs. The unemployment rate in Northern Ireland was such that it would be politically unwise for a government-owned factory like Shorts to announce lay offs. Boredom caused me to seek employment elsewhere. I applied to the Air Registration Board in London for a position in their Flight Test Department and was successful.

CHAPTER 9
AIR REGISTRATION BOARD (1956-1957)

The Air Registration Board was modeled after Lloyds Insurance Company who had experience certifying ships and employed surveyors to inspect ships for seaworthiness.

My interview went well and to my surprise, I was sent to a location in Mayfair for a second interview. A butler ushered me to a large room that was referred to as the clubhouse. Here, I was introduced to Lord Brabazon of Tara, who was a patron of the Air Registration Board.

Brabazon was the first pilot to receive a pilots' license issued by the Royal Aero Club. This he obtained in 1909, when he flew the Shorts built Wright Flyer. At one time he was Minister of Aviation and also headed the Brabazon committee to steer the direction of the aircraft industry.

Named after him, the Brabazon aircraft was built by the Bristol Aircraft Company. It first flew in 1949 and was as big as the Jumbo Jets, but only carried a hundred passengers in lavish luxury. Built to replicate the equivalent comfort of an ocean liner, it offered sleeping accommodation, dining rooms, a promenade, a bar and cinema. With a clientele of only senior civil servants and business executives, it was doomed to be uneconomical and the project was abandoned in 1953 after 400 hours of flight testing.

He asked the butler to have us served tea and cakes. I sank down into a large soft leather couch chair next to this distinguished looking gentleman. We had a pleasant conversation, none of which was associated with airplanes. When tea and cakes were consumed, we both struggled to get out of those chairs, his struggle, being an older man was greater than mine. The interview was over and I left the Lord, wondering why I was there. A few days later, I was offered a post as a Flight Test Surveyor. The good lord must have given me his blessing. Was it because I didn't scratch myself in my nether regions or spill my tea?

Being the new employee, I was given a variety of miscellaneous tasks such as the annual Certificate of Airworthiness renewal tests on a variety of aircraft, like the Douglas Dakota Mark 3, the Handley Page Hermes Mark 4, the Lockheed Constellation, the Boeing Stratocruiser, and the Avro Tudor Mark 4B.

Other tests were to check out various modifications a single engine Auster, a de Havilland Dragon Rapide (A twin engine by-plane), a Douglas Dakota and a Scottish Aviation Twin Pioneer (a single engine short take off and landing aircraft). Many modifications such as the installation of a new type of engine, a new type of propeller and drag improvements, required testing the climb performance, take off performance and stall speed measurements of many modified aircraft.

I participated in many miscellaneous aircraft tests representing the Air Registration Board as a test witness. Company pilots who were not qualified test pilots conducted the tests. They would often get upset when I demanded repeat tests when test conditions were not satisfactory or when test results did not match the expected flight manual data.

The treatment of second officers (co-pilots) by some senior pilots surprised me. It reminded me of the cast system in India, when members of a higher cast would treat the lower cast with disdain. It did not surprise me that accidents occurred due to poor cockpit communication. I am glad to say that current crew training now stresses the importance of mutual respect between members of the crew.

BLACKBURN BEVERLEY

My next assignment was a good one. I was to represent the Board during tropical trials on the Blackburn Beverly Military Transport airplane. The trials were to be conducted in Tripoli, Libya in North Africa.

The Beverley, built by the Blackburn and General Aircraft Company, was a heavy lift military transport aircraft, with four Bristol Centaurus engines. It had an enormous box like body with a large tail boom. Easily removal clamshell doors covered the rear box fuselage. With the doors removed, the aircraft could be used for supply dropping up to 16 ton, such as army battle tanks. Paratroop dropping could also be conducted with 40 paratroops in the freight bay and an additional 30 in the tail boom. The aircraft was also capable of carrying bulky equipment such as oil drilling equipment, which could stick out in the breeze beyond the open rear area.

My role in the test program was to witness aircraft performance, engine and system tests. Tests were to be conducted with and without the clam shell doors installed. These large rear entry doors allowed high volume freight such as army battle tanks to be loaded. When the doors are removed, freight such as oil rig components can be loaded with sections trailing behind in the breeze.

I met up with the fourteen men test crew in Boscombe Down. The test team consisted of a company test pilot, an RAF test pilot, a navigator, 3 flight test engineers, 2 instrumentation technicians, 5 maintenance personnel and a Bristol engine representative.

The aircraft arrived in Libya at Idris, which is the airport near Tripoli, after an overnight stop in Marseille in France. A short drive from the airport was Castle Benito, where we were to stay in Benito Mussolini's old headquarters in Libya. He certainly lived in style and our accommodation was five star. Our rooms had high ceilings, marble floors and were furnished lavishly. There were well-tended gardens and rows of stately palm trees surrounding the buildings. We were by no means, roughing it. I felt like a duke and imagined myself as Il Duce as he was known, strutting about barking out orders. My imagination faded when I remembered his body, after being shot and strung by his feet, was hung for public display on a courtyard in Milan Italy.

The ground rules for the hot weather tests were, that the tests would only be undertaken when the temperature was above 40 deg C (104 deg F). There were many days when the temperature was around 95 deg F, leaving all of us hanging around, checking weather forecasts. On one of those waiting days, a few of us decided to visit the small Town of El-Azizia, which was about one hundred miles away. El-Azizia had two claims to fame. One claim was that it once held the highest recorded temperature of 62.7 deg C (145 deg F). The other claim was that due to the excessive heat, many there, lived in underground caves. Four of us set off early one morning in a Morris Minor rented car.

En route, we explored an abandoned building, which, from the slogans written on walls, was a prison, housing British prisoners. There were many names and hometowns written as well as derisive slogans featuring Hitler and Mussolini. By the time it was noon it was getting very hot and suddenly the car engine stalled. Luckily, we had the Bristol engine rep with us and he came to the conclusion that we had a fuel vapor lock, due to the high temperatures. By the time the fuel line was bled, to get rid of the air lock, it got so hot that even with our shoes on, it was too hot to stand on the pavement. The car body was too hot to touch.

The engine started and we went on our way. Arriving in El-Azizia, we were hot and thirsty and came upon an open air store which supplied cold drinks. No sooner had we left the vehicle to get our much needed drinks, when the store owner rushed out and embraced the engine representative as though he was a long lost friend. He spoke pretty good English and it turns out that he had mistaken the rep for a colonel to whom he served as a batman during the war. He thought that his colonel had come a long way to pay him a visit. Our engine representative did have a distinguished look and sported a fine mustache.

Once the mistaken identity was sorted out, we got our drinks. We asked the store owner where we could find the underground caves and low and behold, he told us that he lived in one and would be honored if we would visit his home. We said we would love to, so he closed up his store and took us to his home. It was a short walk and soon we came across some old ruins. Here we entered a steep tunnel, which descended into a large open courtyard approximately sixty feet below the surface and open to the sky. Surrounding this courtyard were several caves in each of which he kept one of his several wives. The main cave was the living area, which was richly carpeted and strewn with ornate bolster pillows. Reclining on these pillows, we were served strong coffee and some boiled eggs by one of his wives who was well hidden behind her purdah. In the courtyard I could see large thin slices of meat hanging on lines strewed across the yard to dry. On the cave walls were many religious pictures one of which was Jesus and Mary. On asking the significance of this picture, I was made to realize for the first time that Jesus was one of the prophets in the Muslim religion. While we were deep in conversation, I noticed a young woman in a cave across from me. She removed her yashmak (face veil) and smiled at me. Her smiling eyes captivated my attention; the more I stared the sexier the eyes became. Her come-hither look was overwhelming; it was almost like having sex remotely. My sense of decency forced me to take my eyes away from that hypnotic stare. Still occasionally, I would glance over for a cheap thrill. My father had often mentioned the come hither looks he witnessed during his military campaigns in the Arab world. I did not comprehend his statement until I had witnessed it for myself.

Before leaving we were each given bracelets to take home to our wives. I passed by the cave, which housed the smiling beauty and nodded to her. The nod was to show my appreciation of her interest and to convey, "Parting was such sweet sorrow". The hospitality we received was overwhelming it was an experience never to be forgotten. Always,

I will remember those big inviting eyes, framed with a dark tint and long eyelashes, shining like pools of liquid moonlight.

When temperatures were high enough, the tests continued. During lowlevel engine cooling tests with the rear clamshell doors off, I was up in the tail boom where the instrumentation station was located. When all recorded temperatures had reached their peak and stabilized, the test was over and I clamored down the open fuselage and sat on a rear-facing seat to enjoy the expansive open-air view. Only a slim rope was placed across the vast open end as a safety precaution and did not detract from a panoramic view. We were still flying very low over the desert when suddenly I spotted a lone horseman on a beautiful black horse. The horse reared up, frightened by our noise and the rider appeared to have his hands full maintaining control. Soon, he became a speck in the distance. There was nothing but empty desert for miles and miles. I often wondered who that rider was and where he came from.

When the aircraft was not conducting tests relative to civil certification, I was invited to take part and helped out with data reduction and analysis. This made me feel less of an auditing snoop and I welcomed the opportunity and the experience.

The Italian influence in Tripoli was very evident. There were many excellent Italian restaurants. One favorite had a beautiful courtyard with grapevines overhead on a wooden trellis and bunches of grapes hanging down within easy grasp. Ornate statutes, flower gardens and colourful tile floors completed the decor.

The service was as elegant as the decor and the food was as excellent as the surroundings.

On some of the side streets in Tripoli, I noted Libyans and Italians living side by side with children of both races playing happily together. It was a good example of integration that is not always seen elsewhere. With the tests completed, we returned and spent a couple of days in Malta en route home. An evening spent in the capital, Valletta was intriguing. Late in the evening on most nights, the main street was devoid of all traffic and crowds of people dresses in their finery paraded the street up and down sometimes as many as twenty abreast. They all seemed to be in good spirits and often broke up in laughter and song. All the stores were open from which we bought a few trinkets to take home.

Home had been in my mother's place in New Malden Surrey, where my wife and two daughters were staying while I was away. We bought a large Caravan (house trailer) and located it in a beautiful trailer park near Old Welland, where we were now living.

While I was away in Libya, the office was relocated to Chancery Lane in London. We were given higher value lunch tickets due to the higher cost restaurants in the area. It was a nice perk. I frequented an eating-place near the Old Bailey and often would go to the visitor's gallery in that renowned courthouse. It was hard to tear myself away from interesting cases of murder and divorce.

With reorganization at the new location, I was transferred to a department responsible for pilot certification. I was tasked with producing a Flight Manual for an imaginary aircraft, and setting examination papers to evaluate pilots as to their understanding of aircraft systems and use of this Flight Manual. After conducting several examinations and failing quite a few pilots, I realized that an examiners job was not for me. Spending all day in a classroom, as compared with the excitement and travel associated with flight tests, was a backward step. I had heard that Shorts had been given a lot of new contracts and promptly called them on the phone. They welcomed me back and shortly thereafter I was packed and ready to go with a not very mobile home in the freight hold of a ferry boat from Liverpool to Belfast.

Short Seamew with wings folded.

Short SC 1 Vertical Take Off and Landing Research A/c

Adventure in the Air — 67

Short SC 1 with Rolls Royce Flying Beadstead

Short SB 5 Swept Wing Research A/c

Author (left) and Alex Roberts beside U 10

Author (left) and Dickey Turley-George beside PR 9

Adventure in the Air — 69

Canberra U 10 Drone A/c

Canberra PR 9 Spy Plane.

Author in PR 9 new hinged nose and ejector seat.

CHAPTER 10
Return to Shorts in Belfast (1957-1963)

With my family aboard, it was, as usual, a night sea crossing from Liverpool to Belfast with an angry sea. The two halves of our caravan home were in the forward hold and plainly visible as one half was sticking up above the top of the hatch. The waves caused water to splash into the open hold and I had visions of the morning headline news announcing "600 lost at sea as a ship sinks, due to an open hatch". Well! That was not to be, as the crew hastily rigged a large tarpaulin to cover that open hatch.

We arrived after a sleepless night except both my daughters slept throughout the night. Cliff McKee met us. He was a friend and colleague during my previous days at Shorts. Cliff was one of those who I have described elsewhere as being a key ingredient that makes a department successful. We worked well together analyzing data from tests, using slide rules to calculate and prepare plots presenting, stability and control as well as drag and aircraft performance. By working closely with Cliff, I gained much of his aerodynamic knowledge There were no caravan parks near Belfast and Cliff kindly drove me around local farms to get permission to park. Eventually we found a location about a mile from the village of Carnmony. It was a sheep pen near the entrance to a farmhouse. This was to be my home for a while.

It felt great to be back and my colleagues gave me a memorable welcome back party. Then it was back to work.

Initially, I renewed my production flying experience by taking part in pre delivery testing on Canberra B2 and B8 as well as the Seamew, which I had help develop during my previous employment at Shorts. All the 24 Seamews built, were flown to Lossymouth in Scotland where they were cocooned and never flew again. Yes, it was a waste of all our hard work testing that aircraft.

My next assignment was the Canberra U Mk 10 (1957-1958). The Aircraft was to be used for target practice by a variety of ground to air weapons. I was assigned to this project. Short Bros was contracted to develop an unmanned version of the Canberra Bomber, which was to be designated as the Drone Canberra U-10. The aircraft when fully developed would be controlled from an operator's console on the ground.

This consul was duplicated at the test engineer's position, which was the navigator's station. The consol aboard the aircraft had additional hardware, which would allow the engineer to adjust a variety of signals going to the automatic flight control system. Once the entire signal strengths for pitch, roll and yaw and auto throttle, were finalized, all flight operations could be transferred to ground control.

Getting to the stage when ground control was possible was no easy task, but quite exhilarating.

The ground control console, later to be used by a ground operator, was located in front and above me. Banks of switches were available to control the operation of the aircraft. The switches were labeled as follows: - Take off- Climb- Low speed cruise- High speed cruise- Slow turn port- Slow turn starboard- Fast turn port- Fast turn starboard- Slow descent- Fast descent- Land.

Switching any one of these switches changed the attitude of the aircraft, by changing the position of a gyro platform. The switch demanded a speed, which was maintained by automatic throttle control also when selecting a high-speed descent; the switch would also trigger the operation of dive brakes to prevent unacceptable speed increase.

On a landing, a touchdown weight switch signal triggered the release of a tail parachute, which combined with a wheel brake signal, would bring the aircraft to a stop. The drag chute was housed in a fiberglass tail cone, which, when released, fell on to the runway. In order to ascertain the extent of damage to the tail cone, which could affect its reuse, I volunteered to conduct a simple test. Using the company limousine, with its rear trunk cover removed, I strapped myself in the back of the vehicle so that I was able to stand up and drop the cone from the same height, as it would have been released from the aircraft tail.

The test started, the sleek black shining limousine, complete with a uniformed peak capped company driver, drove down the runway and was to toot his horn as soon as speed reached one hundred and ten miles per hour. I heard the horn toot and dropped the tail cone. That part went well but alas, it was poor test planning. When attempting to stop, the brakes started to fade, and we ended up past the end of the runway in the same rocky area that, years before, the Seamew aircraft was damaged when undershooting the runway. I was rudely tossed around and glad I was well strapped in the back of the car, which required new shock absorbers and some bodywork. Later we completed the tests on a longer runway using a different vehicle.

During test flights, I was unable to reach the control switches on the drone U 10, while strapped firmly in my ejector seat. In order to reach, I armed myself with a pair of tongs, which I obtained from the chemistry lab. The tongs had rubber tips, which gave me a good grip on a selected switch. These tongs proved extremely useful during high-speed descents, when I could open the sick port of my pressure helmet squeeze my nose and blow to clear a pressure change in my eardrums.

The tongs also relieved any facial itch, which could not be relieved due to the fixed visor on the helmet. On a long flight it is surprising how irritating an itch could be especially if the area cannot be reached. The tongs became a welcome piece of equipment to the few of us who were fitted with the fixed visor helmet.

Each switch selection initially caused a variety of responses from the two man crew. Comments such as "Did anything happen?" to either " Wow", "Whoops", "Bloody hell" or "Oh-my-God"

A description of an "Oh-my God" switching response is warranted. On first attempting a high altitude transition from a low speed cruise to a high speed descent, all hell broke loose the moment I selected the HS descent switch. The pilots control column rammed forward to it's stop, my stopwatch hit the roof and smashed, papers clipped to my flight board were scattered, the pilot's hands came off the wheel and I believe we both may have lost consciousness temporally. Data recording subsequently revealed that we had reached a peak negative "g" just over minus four, which would account for my smashed stopwatch and his hands up in the air. The negative "g" encountered had the potential for causing "red-out" which is caused by blood rushing up to the head, causing unconsciousness. The dive brakes, which should have automatically deployed on a high speed descent command, did not. The aircraft went completely out of control and recovered mainly by itself after losing approximately ten thousand feet.

A description of a "Whoops" switching response is as follows: We were conducting automatic landing tests for the first time, at a remote airfield chosen because of its long runway. The first attempt was a disaster, resulting in a heavy landing. Immediately after landing, the aircraft taxied at high speed, went off the taxiway onto a road leading to an officers mess, where we were billeted. Here, the pilot jumped out, left the engines running at idle, and ran into the building. I sat in my position wondering what was going on. About twenty minutes later he returned and we continued the landing tests.

All was understood later when he told me that he had soiled his pants after the heavy landing, claiming he had diarrhea. I believe the landing can qualify for a revised response category, changing "Whoops" to "Oh shit".

When commanded by switches, the aircraft responded unlike the normal pilot induced maneuver response, it would respond instantly in a jerky manner, rapidly changing to a new command. For example, on selecting climb from a cruise switch position- bang! And you were immediately in a climb.

When all the tests were finally completed, the aircraft was sent to the weapons test facility in Woomera, Australia, where it was promptly shot down on its second flight by a surface to air missile. The proximity fuse that should have destroyed the missile prior to contact, failed and the aircraft received a direct hit. All that sweat and hard work literall went up in smoke.

Vertical Take Off and Landing Research (Spring 1960)

My next assignment was to direct the research test program on the vertical take off and landing (VTOL) Short SC 1 aircraft, which was to take place at the Royal Aircraft Establishment in Bedford England.

The scope of the test program was limited to the achievement of transition from a vertical take off to horizontal flight and horizontal flight back to a vertical landing. After satisfactory achievement of the transition the aircraft was to be handed over to The Ministry of Technology for further assessment.

The SC 1 aircraft was the first British, fixed wing vertical take off and landing aircraft. It was a single seat delta wing aircraft. Four Rolls Royce RB 108 engines, installed in the mid fuselage behind the pilot, provided vertical thrust and could be tilted slightly fore and aft. In the rear fuselage, a single RB 108 provided forward thrust. The four lift engines, provided a total of 8,600 lbs of thrust. To ensure vertical lift off, lifting thrust had to overcome the total weight of the aircraft. Approximately only 70 gallons of fuel was all that could be carried on board, to ensure lift off. A few extra gallons of fuel, was added, to be used during start up and pre flight checks. A zero speed rocket ejection seat was installed for pilot safety.

Due to the fuel limitation, flight time was restricted to a very short duration, not much more than a single circuit of the airfield. Careful planning was a must to get the most out of each and every flight.

The conventional elevator, aileron and rudder control surfaces cannot provide control during hover and very low speeds. Control under these conditions is provided by high velocity air obtained from the engine compressors and ejected by nozzles at the extremities of the aircraft.

An extremely complex fly by wire system is used to provide control. The system can provide electric signals to control both nozzles and control surfaces or nozzles only. A conversion to full manual control is available in an emergency. An automatic throttle compensates for lift loss with attitude changes. The complete system is in triplicate for reliability.

Previous to my participation, the aircraft had completed conventional flight tests and had investigated vertical up and down flights while being tethered to a large steel gantry. The floor of the gantry was designed to dissipate the hot jet blast, from the four vertical thrust engines and prevent re-ingestion of hot air into the engine intakes. By constantly hovering over the gantry platform, re-ingestion would cause a significant power loss, preventing vertical lift and sending the aircraft back to the ground. Although limited in scope, tethered gantry tests confirmed that the aircraft was unstable in the hover mode and required the computer generated automatic stabilizer system for safe flight.

The aircraft was shipped to Bedford by sea and road and prepared for flight.

A small team totaling six conducted the trials. The team consisted of three ground crew members, a pilot, Alex Roberts, an instrumentation engineer, Jack Hynes and myself. Alex, Jack and myself were often together and were jokingly referred to as the Short Brothers. The brothers, Oswald, Horace and Eustace were the founders of the Short Brothers Aircraft Company in 1909. We were honored to be associated with their name.

Before writing the test program, I had spent several months studying the aircraft systems and the results from tethered gantry tests and reports from the Rolls Royce flying bedstead, which was a crude contraption aptly named. This gave a good insight into one of the hazards to be avoided. This hazard was the re-ingestion of hot exhaust gasses during take off, causing loss of power and a resultant descent back to the ground. Due to a FOD (Foreign Object Debris) problem, the test site was cleared of all loose stones and debris.

Initial vertical take off, hover and landing tests did not go well until, a no-hesitation liftoff technique was adopted by climbing rapidly out of ground effect. Getting rapidly up to approximately 15 ft was to be the drill.

Next came slow speed forward, sideward and rearward excursions using the tilt of the lift engines. This was followed by small pitch and roll excursions during hover. Small gain changes were incorporated in the auto throttle system to give adequate lift compensation to counteract the increase in wing lift with forward speed.

In flight, lift engine starting was next on the agenda. This would be a prelude to transition from conventional flight to vertical flight.

After a conventional take off using the one thrust engine, the lift engine intake pressures were recorded at a variety of forward speeds. Analysis of the records enabled the recommendations for safe in-flight engine starting air speeds.

Starting four engines, while controlling the aircraft in flight, proved to be a somewhat difficult task that took some time. Because of the short flight duration due to the small amount of fuel available, time was precious.

I wished I could have been up in the cockpit with Alex to help, but there was only one crew seat. My contribution was to provide clear and precise sequence of test direction and call out check functions as required.

Noise levels were extremely high and at times it was hard to communicate with the cockpit. The complex fly by wire system caused cautionary yellow warnings on a few occasions when either one of the three attitude gyros or one of the three rate gyros, went out of tolerance. A red warning demanded immediate action.

After achieving small increments of forward speed the transition from hover to full wing-borne flight was the first goal accomplished. A few days later, the lift engines were started in normal flight and the aircraft then slowed down to a vertical landing. This gave the green light for a vertical take off, transition to normal flight followed by a vertical landing. Soon, the fuel truck was much in demand when VTOL (vertical takeoff and landing) circuits became routine. On one circuit, a red light and an audible horn warning indicated that two out of three rate gyros were out of tolerance. This warning came on while the aircraft was in the process of taking off vertically and had just started forward acceleration. The heavy workload probably caused Alex to react to the red alert by immediately reverting to complete manual control. It was a sight to behold. The aircraft pitched, rolled and yawed in an unstable and apparent uncontrolled manner, until sufficient forward speed was gained, allowing the conventional controls to function. It was apparent to all who witnessed the event, especially to Alex, that manual control was only to be used under dire straits. In retrospect, switching off the rate gyros and relying on the three attitude gyros, would have given a better outcome. Anyway all's well that ends well and manual reversion, due to pilot's skill, proved not to be catastrophic.

Before handing over the aircraft to the Royal Aircraft Establishment for further assessment, we conducted several tests to establish an acceptable range of descent velocity and angle of descent, to ensure a safe envelope that would prevent re-ingestion of hot exhaust gasses.

The RAE, having accepted the aircraft, decided to ground it in order to install telemetry instrumentation. It was a wise move and I wished we had the luxury of instant access to flight data on the ground.

When flying resumed, pilots from the RAE, NASA, and the French Air Force conducted flight evaluation. Some of the pilots had Helicopter experience and they concluded that the aircraft was easier to fly than a helicopter.

After the aircraft was returned to Belfast, a new Company test pilot was assigned to the program. He was undertaking familiarization flights, when the aircraft suddenly ploughed into the ground from a low level exercise. He was killed instantly. Investigation revealed that a set of three gyros had toppled prior to engagement, causing violent signals to the computer generated flight control system, thus causing the crash.

I often wondered, why such an important vertical take off and landing aircraft flight test program, was left to a very small team without much support from company experts. The answer I believe is, that the Ministry of Defense had rejected a bold proposal. A new two-seat Fighter/Bomber was to replace the Canberra Bomber. An interesting proposal was made by Shorts to provide a mobile airborne runway with 56 lift engines. The proposed fighter aircraft would be placed on this mobile platform which would lift off vertically, accelerate, release the aircraft and then return and await launching again to retrieve returning aircraft or launch others. This method would allow aircraft designers to design a super high-speed wing, not encumbered by having to produce a wing that was also used for low speeds required for take off and landing. The high-speed wing would have very little drag, allowing very high speeds. The need for a heavy undercarriage, brakes and associated hydraulics would disappear and the huge weight saving could add significantly to the payload. Rocket power could take this project to the stars. It was not to be. This was too much like science fiction and the proposal fell by the wayside, much to the disappointment of the Company design team.

Short SB5. Research aircraft. (Fall 1960)

The SB5 was designed to investigate slow speed handling qualities of a proposed supersonic fighter aircraft. The wing sweep-back could be adjusted on the ground to 50, 60, and 69 degrees of sweep. The rear fuselage was easily detachable, allowing either a tailplane set on top of the fin (T-tail) or with the tail plane set low, below the fuselage. The rear fuselage housed two brake parachutes and one anti-spin parachute.

In the period 1952 and 1953, the aircraft had investigated the 50 and 60 degrees of wing sweep-back with the high and low tail-plane configuration.

Extensive modifications were incorporated in the aircraft prior to conducting 69 deg sweep tests. The Rolls Royce Derwent engine was replaced with a more powerful Bristol Orpheus for greater thrust. An ejector seat was also installed. It had been previously established that the low tail configuration was superior and this configuration was installed for the 69 deg tests.

I was assigned to direct the tests once again at RAE Bedford. Here, the task was to conduct 69 degrees of wing sweep tests with the low tail plane configuration. This was, at the time, the greatest degree of wingsweep in the world. There was a big concern that due to the high degree of sweep, the small moment arm of the ailerons, would provide insufficient lateral control to pick up the wing when it inadvertently dropped.

The long runway at Bedford was well suited for safe conduct of high-speed taxi and short hops off the runway, prior to the first flight. The morning of the first high speed taxi runs, gave us a cloudless sky and a steady light headwind. These were perfect conditions, but we had no pilot. The pilot assigned for the task had simply disappeared. After three days of searching in Belfast and Bedford, the search ended when he was found to be in a hospital suffering from a nervous breakdown.

The tests commenced after the arrival of a replacement pilot and another fine day. High speed taxi and skips off and on the runway, indeed showed poor lateral control. This was soon found to be no problem at all, as the use of the rudder was found to provided adequate lateral or roll control. This was a characteristic of the highly swept wing.

All the handling quality tests, which included stall tests, were completed in a very short time. Absolutely no problems were encountered. This was the only time that I can remember, after all my experience, before or after, that a test program was completed ahead of schedule without any hitches.

The successful tests, confirmed the wing sweep and low tailplane configuration adopted for the P1 fighter, which was later named the English Electric Lightning.

After completion of all tests, the aircraft was used by the Empire test Pilots School at Boscombe Down, to give budding test pilots experience with swept wing aircraft.

CHAPTER 11
Canberra PR 9 (1958-1963)

The English Electric Canberra PR 9 was developed and manufactured by Short Bros and Harland in Belfast Northern Ireland under contract by the English Electric Aircraft Company.

The cold war demanded the introduction of high altitude spy planes, thus the PR 9 came into being. The wingspan and wing area were increased over the standard Canberra. In July 1955 The English electric company, using a modified Canberra B 8 aircraft, conducted initial tests on this new wing. Initial handling tests were satisfactory, so the decision to build and test the prototype PR 9 in Belfast was made.

XH 129, the first prototype, flew in July 1958. For most of the development flying on the PR 9, Alex Roberts, a young pilot straight out of the Royal Air Force, accompanied me. He was just twenty-one years old, enthusiastic and competent with maturity beyond his age. I believe we made a good team.

Prior to achieving the delivery stage the PR 9 test program encountered a fair amount of set backs.

Early in the test program the aircraft encountered a fatal incident.

As per the contract, after 25 hours of initial testing at our home base in Belfast, the aircraft was flown to Wharton in Lancashire, where the English Electric Flight Test Center was located. Here, a short evaluation by their company pilots and flight test engineers was to take place. A couple of low-level assessments were conducted, including one by the Chief Test Pilot, whose only statement was "You chaps from Belfast have built a fine aircraft."

When it came time for a high altitude assessment, their young test engineer did not have his pressure suit modified with a shut off valve to protect him in the unlikely event that he was required to bail out at high altitude. This left me as the only person with the correct equipment.

I was scheduled to fly early in the morning with one of the English Electric test pilots. The morning of the test, the aircraft developed a snag that needed to be fixed, so the flight was delayed. Meanwhile, the young test engineer as keen as mustard, drove fifty miles to the company that manufactured the missing shut off valve. The valve was fitted to his suit and he got back to the airfield after lunch and pleaded with me to let him take my place. I briefed him on the test program, which had changed to include a maximum speed/'g' envelope demonstration at low level. I helped strap him to his seat, and reminded him of the bailout technique which was to jettison the entrance door, get down to his knees and roll out.

Note: the navigators/test engineers seat in this aircraft was not an ejector seat.

The flight took off and about fifteen minutes later, when back in the flight test office, I heard the news; the aircraft had crashed at sea, In front of the crowded promenade at the Blackpool beach.

The aircraft had started the tests by flying at low level at maximum speed and was in a tight turn attempting to pull maximum "g," when the complete wing on one side snapped off. The pilot just had time to say goodbye as he used his ejector seat to successfully bail out at very low level. Divers later found the body of the test engineer still strapped to his seat. Knowing that this spectacular crash, being viewed by thousands on the beach, would be all over the press, I called home to reassure my family that I was not involved in that accident. The Gods were with me again.

It did not take long; the replacement test aircraft was fitted with test instruments and was ready to conduct the investigation as to the cause of the wing failure.

Wing leading edge attachment brackets were the main suspects and strengthened and modified brackets on the port and starboard wing were instrumented with load measuring strain gauges.

The amplifiers used on the strain gauges were notorious for drifting with time and also with temperature changes affected the output of the strain gauges. Armed with a never-to-exceed value, we commenced the tests.

Before each test point, the aircraft was put into a parabolic "pushover" maneuver to obtain zero "g" and unload the bracket to give a current zero load measurement on the bracket. The maneuver was similar to the one used to familiarize astronauts with zero "g." Each zero "g "maneuver, was immediately followed, by flying at maximum speed and loading the wing brackets by increasing "g" loads.

The difference between the zero loads and the test load was obtained and plotted on-board against increasing "g" values. The values obtained would give confidence in advancing to the next "g" increase.

Before advancing to the maximum "g," as previously agreed, the aircraft returned to base for careful analysis of data. The analysis revealed that loads obtained were close to the maximum load on the previous bracket that had failed with such catastrophic results. The data gathered showed that there was enough of a safety margin in hand with the new bracket. Since my on board data analysis closely resembled the carefully analyzed data, the decision to go ahead to max speed and max "g" was given.

I was glad to say we successfully demonstrated the maximum speed and "g" envelope of the aircraft. The wing stayed on and I sure hoped It would. With the problem solved, tests continued on the second aircraft and soon we were to encounter another potential disaster.

Just after taking off to conduct fuel system and heating and ventilation tests, the control tower reported seeing a white vapor trail coming from the starboard engine area. We then flew directly over the tower and they reported that the trail was no longer there. As a precaution, it was decided to land and have the aircraft inspected. As per standard practice, the fuel boost pumps were switched on prior to landing to ensure adequate fuel supply if high power was to be used on a go around.

During the approach, the tower reported a large vapor trail had reappeared.

We were on final approach about fifteen feet above the runway at approximately one hundred and ten knots when a muffled explosion occurred on the starboard side. The next thing I heard over the intercom, was Alex saying "Christ its on fire," soon followed with "I can't hold it." Then I lost communication. A new multi-service connector had been introduced to save time during "strapping in." All aircraft service lines combine at this point and are connected by one single action, connecting oxygen, anti -g, air ventilation supply, radio and intercom. If disconnected, the lines would be sealed and emergency oxygen contained in the pressure suit will be supplied. This connecter came unlocked. This had happened quite often and a fix was in the works. I could see the pilot's feet, indicating that he had not yet ejected. I tried to make my way to jettison the exit hatch, but an invisible hand prevented me from extracting myself away from the seat. I lunged rapidly forward to no avail. Ever since the loss of the first aircraft, I had unwisely decided not to strap myself in my seat using the blue seat straps.

When I hurriedly decided to leave my seat, I soon discovered that my parachute straps had been entangled with the seat straps, which were behind me. The only way to get to the door was to unfasten my parachute, disentangle the parachute straps from the seat straps, don the parachute again and head to the door. Since I had no other choice, I took my chute off untangled it and climbed back into my parachute harness.

This was not an easy task in a confined area. All this time I was watching the pilot's feet to see if he was still there. His flying boots were the only view of him I had from my position. Smoke was coming out of a heating duct, which was located under my navigator's table and it was getting hot. I envisioned flames surrounding the fuselage. All this took quite a bit of time and assuming my end was nigh, I did not see my life flash by, as folklaw predicted for such impending doom circumstances.

All I remember saying to myself was "Rose'Meyer, you have got yourself into an awful mess." Alex was still there, at least his boots were and the aircraft seemed to be back in control. I went back to my seat and re-established communication. He was glad to know that I was still on board, as not hearing from me, he had assumed that I had abandoned ship at a very low level with little chance of survival. He then went on to inform me that the only reason he did not eject was that the force required to hold the wings somewhat level was so great that if he let go of the control wheel, to pull the ejector seat handle located above his head, he would have rolled over and ejected into the ground. Note: Most of all this happened less than a hundred feet above the ground.

We decided to land at a military airfield nearby which had a long runway. We came in fast on one engine at a speed well above normal landing speeds and made several attempts to land, finally making it on the third attempt.

Many eyewitnesses were watching from the windows of the flight test office and stated that they had witnessed a perfect demonstration Of an aircraft, out of control, during an attempted go-around. Apparently we went between two hangers, missed a ship in Belfast harbor and disappeared into a valley between two hills. They were awaiting a puff of smoke indicating, we had crashed.

For those readers who have flown multi engine aircraft and are aware of the minimum take off control speed (Vmca), be it known that the minimum Vmca for the Canberra PR 9 is 190 knots and the go around was attempted at 110 knots. This was not a survivable speed and yet we survived. Cliff McKee, a colleague and friend of mine who lived near my home, later stated that he was wondering what words he would use to inform my family of my death.

Once again, the Gods were with us.

The culprit that caused this episode was a flexible fuel connector in the engine bay located near the jet pipe. This connected the fuel in the fuel collector box, which housed the fuel boost pumps, to the engine.

With boost pumps selected on, the pressure in the fuel pipes increased, causing the connector to come loose and spray the hot jet pipe with fuel.

This caused the explosion. The sudden explosion actually extinguished the flames al la the technique used to extinguish oil rig fires. Smoke and heat I had seen and felt was no more than the results of selecting full heat on the new heating system and as the duct pipes got hot, residual oil (maybe due to greasy hands during installation) burnt off the surface of the duct pipes. This was a common occurrence on initial heating check out on production aircraft. Under the circumstances, I did not even consider that factor and let my imagination run riot.

After landing safely, we shared a taxi back to the flight test office, got dressed and went our separate ways. My car would not start, so I walked a good mile to catch a bus to the Smithfield bus station where I waited a long time for a bus to Carnmony. After a two-mile walk from the bus stop in Carnmony, cold and soaking wet from the rain, I arrived home, better late than never. While sitting in the Smithfield bus station, I contemplated on the war years, when aircrew would return from missions over enemy territory, after being shot at by fighter aircraft or surrounded by the telltale puffs of smoke from anti-aircraft guns. Their hearts would be beating with excitement and apprehension of the next mission, as did mine. I could clearly understand the thoughts of so many gallant airmen, as they came back to tranquil surroundings. No one could possibly imagine what the wartime crew had been subjected to?

It was like a calm after an almighty storm. About a month later, we were conducting a series of drag measurements and were flying at a variety of steady speeds at 20,000 ft when I noticed that the speed was far from steady. When I heard Alex burst into song and the aircraft maneuvering to his tune, I realized he was lacking oxygen. He initially paid no attention to the suggestion that he descend to a lower altitude, until I changed the next test point to 5,000 ft even though it was to be at

40,000 ft. Once back at 5,000 ft, we mutually decided to head back and have his oxygen equipment replaced. Did the Gods once again smile on me?

Soon after this incident, we were asked to ferry the aircraft to Boscome Down. Here the aircraft was put in a hangar and I was briefed on the task I was required to undertake. A demonstration was to take place, which would be witnessed by a number of high-ranking officers. The demonstration was to show the ease (or lack of) of exiting the aircraft in an emergency. I understood that this demonstration was the result of objections by a group of navigators who were rightfully upset that both crew members of the PR 9 aircraft were not given equal escape Facilities, i.e., the navigator should have an ejector seat, as did the pilot.

Two of us were chosen to conduct the demonstration: A Royal Air Force navigator who was a foot taller than I and weighed almost twice as much and myself. As instructed, I laid, face down in the bomb aimer's position, forward in the nose, where the bomb aimer could look through a bombsight mounted in the Perspex nose cone. My pressure suit was then inflated and upon a signal, which was a bang on the side of the fuselage, I was to get back to the exit door, about eight feet behind, and jettison it. My time to reach the door and jettison the door was just less than sixteen seconds. I did not know what the criteria for time was, I did my best and it still took around sixteen seconds. Maneuvering around in an inflated pressure suit was not easy.

The Air Force navigator took more than two minutes to complete the same demonstration. Why it took so long I will never know. Perhaps a deliberate slow down helped the navigators cause. The weather was deteriorating, so we departed for Belfast in a hurry and I was not able to attend a formal debriefing.

A short while later we were given the go ahead to design and install a navigators ejector seat system for the Canberra P R 9 aircraft.

The company's airfield was located in Sydenham, near the heart of Belfast. One runway ended up a few hundred feet from the water's edge in Belfast Lough. Seagulls were a real hazard, during take offs. They would gather on the runway, feasting on worms. Several methods were applied to attempt to get rid of them from random gunshots to employing a falconer to use several birds to scare the gulls. The falconer did a better job than the gunfire but still the seagulls remained.

During a take off to conduct engine-handling tests, we encountered a flock of seagulls and the pilot noticed a slight surge or hiccup on the

Port engine. This only lasted a moment after which the engine appeared perfectly fine. We carried on and completed the tests, which, included slam accelerations at heights from low level to sixty-five thousand feet.

These accelerations were required to check for any tendency for the engines to encounter a compressor surge and possibly fame-out and shut down. All tests went off without a hitch. On my instrument panel, I had noticed a slightly higher Jet-pipe temperature on the port engine but thought nothing of it. After landing a small group of ground personnel gathered around the port engine, shaking their heads. As we climbed out of the aircraft, we saw a very large seagull with its wings spread over a good one third of the engine intake. The bird was dead but looked in surprisingly good shape. The Rolls Royce Avon engines in the aircraft were always reliable and after this incident, I had an even greater respect for their design and reliability.

A few days after the seagull incident, the gods came back to watch over me. We were returning from a test flight and were a couple of thousand feet high and near the airfield when suddenly both engines flamed out and the aircraft became a glider. We were at the right place at a bad time and with great piloting skill, Dickey Turley George brought me home safely. Once again the Gods were with us.

This incident was no fault of those magnificent Rolls Royce engines that had not given us a jot of problems. What had happened was that the fuel collector tank had become clogged with sheets of rubberized sealant, which had been sprayed on the inside the fuel tanks during the manufacture of the tanks. This thick rubber coating was meant to protect the fuel tanks from small arms fire or small shrapnel by enabling the flexible material to return to near its original shape. The production process on a batch of these tanks had not been adhered to and the cleaning and de-greasing of the tank inside surfaces had not been done. The fuel boost pumps were completely clogged with sheet sized pieces of rubber material. It is always amazing that a small mistake can lead to serious consequences.

We had completed autopilot tests and the autopilot became a useful tool when conducting climb and cruise tests. It is always essential to disengage the autopilot when taking off and landing. On one occasion, the autopilot would not disengage when selected off, prior to landing.

Even pulling the autopilot circuit breaker did not help. This was quite a mystery. The pilot had to use a lot of muscle power to fight the autopilot by overcoming the autopilot servo loads. Thanks to good piloting skills, we landed safely. Electricians and specialists examined the autopilot and could find nothing wrong. One smart electrician asked me what switches had been selected at the navigators position when the incident occurred. With the aircraft in the hangar, I turned on all the usual switches. One of the switches was an IFF (Identification Friend or Foe) switch. There was not yet an IFF black box installed in the aircraft.

This was a secret device which would be installed after delivery to the Air Force. Nevertheless, I had got in the habit of turning this switch on after start up, in conjunction with many other switches. As soon as I turned this switch on, low and behold, the autopilot would not disengage.

We could now reproduce the problem. It turns out that a small piece of metal filing was lodged between two pins in a connector, which provided the power supply to both the autopilot and the IFF circuit.

When turning the IFF switch on, the stray filing connected electrical power to the autopilot circuit. It was not good engineering practice to have two different power supplies in a single connector. Imagine that a tiny, stray piece of metal could have caused such a serious situation.

With fuel tip-tanks installed on the high altitude PR 9, aircraft handling qualities had to be evaluated. On the very first take off with the tanks, we hit a flock of seagulls and severely damaged the port fuel tip tank. A replacement tank was soon fitted and the test program continued.

Low speed handling, including stall tests, were quite satisfactory.

High-speed investigation showed a distinct reduction in the limiting Mach Number. Note! Mach 1 is the speed of sound. (760 mph at sea level) Severe buffeting characteristics had occurred at .85 Mach without the tanks and at .79 Mach with the tanks installed. As we climbed to high altitude around 64,000 feet, we came close to the phenomena known as "Coffin Corner". There is very little air at these high altitudes and the rarefied air made it very difficult to descend to a lower altitude.

There was now a very small margin between maximum speed and minimum speed. A few knots separated the aircraft from either stalling or shaking violently as it exceeded its maximum speed or Mach Number limit.

Only gentle maneuvering was tolerated as the aircraft descended slowly with a very narrow speed margin to play with. The famous Lockheed U 2 spy plane that was shot down by a surface to air missile over the Russia and sparked an international crisis, had similar encounters with coffin corner and because of the difficulty it encountered flying at altitude, was nicknamed the Dragon Lady. The term, coffin corner, is aptly named. When an aircraft reaches that dangerous corner, it cannot decrease speed or increase speed. Maneuvering is almost impossible. The difficulty in controlling the aircraft is made apparent to the pilot before reaching that awful corner. Perhaps the Gods are providing a warning, telling the pilot to cease climbing up any further toward the heavens.

Flying at extremely high altitudes is not a task for the faint hearted. The Royal Air Force had encountered several fatal occurrences when pilots in fighter aircraft, flying around 45,000 ft ejected their overhead canopy, prior to attempting to use the ejector seat to bail out. In the wreckage, they were found to be still strapped to their seat. Since the P R 9 had a fighter type cockpit, with an overhead canopy capable of being ejected, we were asked to conduct the accident investigation.

Note that fighter pilots were not equipped with pressure suits as these were not required at heights below 48,000 ft.

Twenty probes were installed to measure both air pressure and temperature at locations near the aircrew body locations and also near the oxygen regulators. This data was recorded at the navigator's table.

The pilot's overhead canopy was removed before flight. Because of the expected cold air blast in the cockpit area and the danger of frostbite, the pilot, Alex Roberts, was equipped with an ancient leather face mask, which had narrow slits for his eyes. Where this mask came from I do not know, perhaps it came from an Air Force museum. We both wore extra warm socks and underwear. With his goggles on, Alex looked like Darth Vader. The first test points were at 15,000 ft. The noise level with the canopy removed was not too bad. I was waiting for temperatures and pressure reading to stabilize before calling for an increase in speed. I noticed that the aircraft was wallowing about and was just about to ask what was happening, when Alex asked me if I had my dingy knife handy. I thought his dingy had inadvertently inflated which would not be a good situation in a cramped cockpit. He asked me if I could reach his neck with my knife? After a short pause wondering why he was trying to slit his throat, I realized that he was flying completely blind. The more he tugged at his mask, his narrow eye slits were moved further away from his eyes. I tried to reach and cut the mask strap but my parachute would not allow me to stand erect. With a gaping hole above his head, I was not prepared to remove my parachute to reach him. The aircraft by now had become seriously out of control. Standing half erect and fighting the g forces I was being subjected to, I almost reached his neck and may have given him a few shaving nicks. In desperation, he grabbed the knife and slit the offending strap. To this day, I wonder what could have happened if that knife had slipped from his grasp and dropped out of reach, leaving him blind.

Since the air blast in the cockpit was quite innocuous the mask was discarded and the tests resumed.

Test results showed that the cockpit altitude, on average, increased 6.000 ft above the height that the aircraft had been flying. This was due to the suction created by the airflow over the pilot's windscreen. Pressures at the navigators position, were not too different from that recorded at the cockpit. I do not remember the temperature readings, but considering the open cockpit, we returned not as cold as we expected to be.

Fighter pilots flying at around 45,000 ft would be subjected to an equivalent altitude of 51,000 ft, immediately when the overhead canopy was released. Without pressure suits, instant death would result. Death, caused by the boiling of blood and other body fluids. They died still strapped to their seats, unable to pull the ejector seat handle.

Accident inveestigation was now completed and the problem understood.

Doctors from the Institute of Aviation Medicine were concerned with the environment that air crew, wearing pressure suits, would encounter while conducting low level operations on a hot day. An air ventilated suite and a skullcap, had been developed to keep the body and head of pressure-suit-wearing aircrews, cool. When they heard that we were about to conduct hot weather testing in Bahrain in the Persian Gulf, they suggested that they would like to supply us with air cooled suites and would participate by monitoring our vital signs and weight loss during the hot day test program.

Bahrain was chosen as a hot location because of the combination of high temperatures and high humidity. An air ventilated supply was readily available in the cockpit area as it was used for demisting the pilot's and bomb aimer's viewing area. The air ventilation supply also provided an important function in the camera bay compartment.

The vital camera compartment had extremely tight requirements with respect to temperature distribution. The temperature at the film magazine location at the top of the camera compartment was not to be more than two degrees, centigrade different from the lens area, which was at the bottom of the compartment. The camera was extremely large and approximately five feet separated the top of the magazine from the camera lens. The photos taken from a high altitude were to be capable of reading a vehicle number plate. We had spent many weeks getting the temperature distribution to meet the tight tolerance requirements. It was also important to know that a rapid climb to altitude, especially under high humidity conditions, would not cause the camera clear-view panel or the pilot's view to mist up.

Considering the additional demand for ventilated air, an additional supply was introduced at the navigator's station. Engine bleed air was mixed with air that passed through an air cooler and the navigator could mix the hot and cold air to his liking.

CHAPTER 12

After completion of the test program and delivery to the Royal Air Force, the PR 9 aircraft were fitted with satellite-communication antenna equipment supplied by the United States, allowing the transmission of photographs in real time to intelligent centers anywhere in the world. The technology used was the same as that used on the Lockheed U 2 Spy plane.

The PR 9's flew deep into the Soviet Union and China during the cold war period.

Off we went to Bahrain, stopping overnight in Khartoum in the Sudan. On getting out of the aircraft in Bahrain, we noticed a large contingent of personnel. Apart from our own instrumentation engineer and ground crews, there were a couple of doctors, an Air Force representative, a civilian air safety officer, a representative from the pressure suit manufacturer and a couple of executives from the English Electric Company.

We were all staying at the British Oversees Airways Corporation Hotel. The next morning, we were fitted with our custom-made air ventilated suits. They were made of nylon with a multitude of air tubes ending at a variety of body locations. Each tube had approximately a one eight-inch diameter hole through which air would circulate. The skullcap was attached to the main body suit by a hose with a tap located at the side of the neck. The tap was only to be opened at low level and not at altitudes where it could mix with the oxygen supply and dilute the intake of oxygen.

The first task was to get weighed, have blood pressure and temperature recorded together with pulse rate and breathing rate, immediately before getting dressed for flight. We were asked to take our body temperature, pulse rate and breathing rate every fifteen minutes during flight. Post flight would require the same data.

Prior to getting dressed, my body temperature was taken and found to be slightly over 100 deg F. This simply put an end to the tests. It was explained to me that often when rapidly moving from a cold to a hot climate, sweat glands could take a while to work efficiently, hence my increase in body temperature. I was feeling perfectly fine but embarrassed at holding up procedures especially with the large number of onlookers present. It took three whole days touring and relaxing before my temperature became normal. I had never known such interest in my health and welfare.

The tests commenced just passed noon and I now realize the advantage current astronauts have when they use light weight air conditioning Hand-held packs to cool their air ventilated suits when on the move.

An air supply was hooked up to the ventilated suites, as we got dressed. This was always quite a big chore. As we disconnected and left the dressing room, it felt quite hot until we got into a transport vehicle where we re-engaged the cooling air supply. Leaving the transport and our cooling air supply and conducting pre-flight checks, entering the aircraft and strapping in prior to engine start, took enough time that I could feel myself sweating profusely. Those sweat glands must have been working well. After connecting my suit air supply and selecting to max cool, I did not note much of a cooling effect and put it down to low engine power. During taxiing to the take off position, I still did not notice much cooling affect. The pilot stated that his cooling felt fine. During take off and immediately after, I felt an immediate cooling sensation as the airflow to my suit increased with engine power. Very soon it became unbearably hot and I turned off the air supply. I started sweating and again turned it on. An immediate cooling sensation was followed by it again becoming unbearably hot. I decided to check the air supply temperature.

It was hard to believe, but the air temperature to cool me was just over 100 deg C, the temperature of boiling water and about the temperature of the hot air from a hair dryer. It appeared that everything was cool until my sweat evaporated. My sweating skin behaved like the kus kus tattie window blinds used in India to cause warm breezes to cool the surface and create a cool interior. As a note, heat transfer in water is significantly greater than transfer of heat by air, which is why I survived this ordeal. The test had to be abandoned until the problem was resolved. On the way back I was constantly turning the "hot" air supply on and off as my sweat glands dictated. One suggestion after landing and discussing the problem was that the hot/cold switch was plumbed wrong and that we should continue the flight by having me select hot instead of cold. I refused to go along with this idea and suggested we took a good look at the plumbing. The plumbing turned out to be correct, so had I selected hot, the temperature would have been above 500 deg C and I would have been toast. Further investigation revealed that the bleed air pipe that had been hurriedly installed was not insulated where it was located around the hot jet pipe.

At high power, heat was being transferred, to the air pipe causing the high temperatures that I felt. With pipe insulation carefully installed, I was cool again.

All low level engine, generator, hydraulic oil, inverter bay, navigation bay and crew cooling tests were completed and soon the aviation doctors left with lots of our body data, to compile their report.

As always, it was a very humid day and a demisting of the crew windscreen and camera lenses were investigated. A rapid climb to altitude was undertaken. The aircraft went from low, hot and humid to high, cold and dry conditions. The temperature at sea level was about + 43 deg C with a humidity around 98%. The temperature at maximum altitude was -59 deg C. Apart from not unusual snowflakes initially coming out of the air vents, all viewing areas were clear. I had waited long enough for temperatures in the camera bay to stabilize and had recorded the final reading, when we both heard the sound of a crack.

We started to descend and I noticed a large crack extending all the way across the transparent nose cone, which was at the bomb aimer's position.

On returning, we put the aircraft into the hangar, which was owned by the Bahrain Oil Company. No sooner had we got into the hangar to start removing the nose cone, than we were told to vacate and park on the hot tarmac outside. The hanger was to be taken over by hundreds of rhesus monkeys. They were the type I had often come across in India.

Their distinguishing features were reddish faces and a red rump. They were on the way to the UK to be used for medical research. The Douglas DC 4 that was en route from India had developed engine trouble and was forced to land in Bahrain. The DC 4 taxied to the front of our hangar and disgorged hundreds of cages each with a very unhappy monkey.

They were all placed in rows and were spread all over the hangar floor. If they had remained outside in the aircraft, they would all have died from the excessive heat, as the fuselage would soon become a hot oven. The office we were using to examine data and write reports was at the back end of the hangar and we had to plough through a very noisy and soon, very smelly cages. As I made eye contact with a quiet and sad looking monkey, I wondered if he had been saved from a hot oven, only to be sent to a worse fate.

Work continued outside in the hot sun. Some shade was provided over the nose section by a hastily rigged tarpaulin. Our hard working mechanic, with his body twisted in unnatural positions, was inside the nose area attempting to undo a large number of fasteners attaching the nose cone to the fuselage. The monkeys may have been safe and somewhat cool in the hangar but our mechanic paid the price for his efforts by suffering heat exhaustion and had to be hospitalized. I remembered suffering from a heat stroke a long time ago when living in New Delhi and knew how serious it was, as heat exhaustion can easily lead to a heat stroke. A few days in the hospital and he was as good as new. A replacement nose cone arrived and was installed. It was suggested that the nose cone cracked because of a combination of uneven stresses during tightening of the attachment fasteners and the resulting thermal stresses associated with rapid changes in temperature.

Soon, all tests were completed and the aircraft was prepared to return home. During our five-week stay in Bahrain, we had been invited to join the Gymkhana Club, where we often went after work and made many friends there. The club at first appeared very stuffy with most members dressed elegantly, ladies in evening gowns and many men in tuxedos. We were at a table right beside the swimming pool. I was dressed in my best bib and tucker (khaki trousers, white shirt and red tie) and was walking by the edge of that brilliantly lit pool, when a member of our group nudged me into the pool. Naturally I retaliated, and in he went. Soon it became a free for all and before long, women in long evening dresses, were pushing men some of whom were wearing tuxedos and men were pushing women into the pool. The scene reminded me of a similar scene in a movie called "The Party," starring Peter Sellers. It Was a night to be remembered.

Many members of the staff at the hotel we were staying in were from India and I was able to converse with some who could speak Hindi. I was amazed that I could recall words that I never knew, I knew. I had complained to them that I did not see any of their great curries on the menu. The next day, they served me a delicious goat curry, which they had prepared for themselves. After that, I was occasionally served some of their own meals, and boy! Did they eat well?

Another place we visited was the Bahrain Oil Company's recreation center, which was located several miles out of town. It was a huge complex sporting an Olympic-size swimming pool. On one end of a huge hangar-type building, was a seemingly endless bar, which was touted as the longest bar in the world. There was a billiard room, which contained at least forty large billiard tables. The table tennis room and the darts room contained a horrendous number of tables and dartboards.

The washrooms rivaled those in five star hotels. With all these facilities and during several visits, we hardly saw anyone around the recreation center, which had been built to encourage employees to remain with the company.

After two of us were swimming in their huge swimming pool, I thought it would be a good idea to swim in the Persian Gulf at the nearby beach area. It was dark and growing darker when I found myself lost wondering which way to swim to get back to the beach. It was a nasty moment, I shouted out but could hear no response. Listening carefully, I thought I could hear the sound of the surf on the beach, but the sound appeared to be coming from several directions. I went towards my best guess and it was a good one. The gods were with me again.

Soon after, the tests were over and we headed home.

On returning to Belfast, I was in for a treat. The improved PR 9, sporting its new hinged nose with a navigator's ejector seat, was ready for its first flight. A frangible fiberglass hatch was located above the ejector seat, which allowed ejection through the hatch without having to eject the hatch. A periscope in front of the seat would allow all around vision, not just forward and down but rearward as well. This periscope was also used for bomb and photograph aiming. On my left, at eye level was a small window. There was no access to the pilot's cockpit, which was behind me. If you were at all claustrophobic, you would know it when the hinged nose used for entering, was clanged shut.

We conducted several flights, checking out system functions and handling qualities. I had always carried an altimeter, which I used to check cabin altitude relative to the aircraft altitude. This I used to place on the navigators table. There was not a table on this aircraft, so I lay the instrument, face up in a slot, which was vacant awaiting the arrival of a radio altimeter (used to measure tapeline height above the ground).

We were over fifty-thousand feet high when I decided to look at my cabin altitude indicator. Imagine my surprise, when, I saw a value close to sea level. I just could not understand it, since it meant that we were dangerously over pressurized. The small cockpit cabin altitude indicator appeared to be normal. The pilot reported that one of his canopy seals was showing a bit of daylight, indicating a pressure leak. I had witnessed many pressurization ground tests when six pounds per square inch was applied to the cabin, which was two pounds per square inch over the standard operating cabin pressure. Safety rules dictated that, only essential ground crew were to be around during the test, and that they were required to stay behind a baffle in case something broke loose, like a window or panel. As we made an emergency descent, I was fully aware that my cabin reading indicated that we had ten pounds per square inch trying to burst the cabin area. I took my cabin altitude indicator out of the slot it was in, shook it and tapped it, when all of a sudden the needle wound up so fast that the needle was a blur. I assumed that the cabin had sprung a major leak and that we had suffered a sudden decompression.

There was no bang, no sudden fog or change in the oxygen regulator supply, which would normally accompany a sudden decompression. After landing, scratching my head, I took the cabin indicator to the instrument laboratory and they found that it was damaged due to the high wind up it had encountered. It soon became evident that there was only an imaginary and not a real emergency. The indicator had been placed in a wooden box which had thick rubber pads stuck on the inside to protect the indicator from damage. A narrow tube, which sensed the air pressure when placed in the position I had placed it in at the start of the flight, had sealed itself against the rubber and trapped the airfield pressure. When I took the box, housing the indicator, out from it's stored location and shook it, the sensing tube and the rubber protection pulled apart, allowing the indicator to suddenly sense the correct cabin altitude and rapidly wind up to the correct pressure.

Smoke in a cockpit is plenty of cause for alarm. We were at high altitude, when the navigation equipment compartment, located in the crew area started smoking. The smoke being pretty thick, we declared an emergency. We isolated the emergency battery, pulled all suspected circuit breakers and descended rapidly.

The closest airfield at the end of the descent was Nutts Corner, which at that time was the civil airport for Belfast. After landing, the only entry into the terminal building appeared to be through the baggage claim area. We both came in, surrounded by a variety of suitcases, on a conveyer belt. In those days and even today it would be unusual for two men dressed in space suits, suddenly appearing with the baggage, in front of a large crowd. We certainly were the strangest baggage they ever saw judging by the expressions on many faces. No one claimed us, especially, when, Dickey Turley George, who was piloting the plane started stripping down to his underwear. I guess it was somewhat warm in that baggage claim's room so, unabashed, he took off his helmet, pressure jacket, flying suit, boots and g-suit, performing a sort of strip tease in front of many somewhat shocked men, women and children on-lookers including myself.

With his flying suit back on, covering up his underwear, we took a taxi back to our home base, after calling for a repair crew to replace the burnt out wires and investigate the cause.

We were getting close to completing the development phase of the PR 9 aircraft program, and I was busy completing final reports for Certificate of Airworthyness, release.

Before heading to Boscombe Down, to obtain radio and navigation signal strengths and also to evaluate the newly installed radio altimeter, we went to a section of the airfield to witness the final clearance of the newly installed navigators' ejector seat. The Martin Baker Company, who manufactured the ejector seat, conducted this test. I remember thinking, that it was a bit late for these tests, as I had been occupying that seat for the last three months. A mock up section of the aircraft forward nose area was placed on its side. The ejector seat and a dummy representation of a navigator were installed, as was the fiberglass hatch.

Three, two, one and the ejector sea, was fired, sending the seat with its occupant, through the hatch and into some bails of hay.

The dummy appeared to have damage to its lower limbs below the knee. We all thought it was caused, by hitting the haystack. The Martin Baker Personnel, went straight to the fiberglass hatch where they had lined the inside frame with plasticine. Here, they noted tell-tale dents, indicating where the knees hit the frame. The normal procedure for ejecting was to pull the ejector handle above one's head; this would pull a sheet of tough material over the face to prevent damage due to a sudden air blast. By reaching above to pull the handle, the spine is placed in the correct position to prevent spinal damage due to sudden upward acceleration as the ejection gun cartridge is fired. At the same time, straps around the ankle would tighten and secure the legs to prevent them from flaying about. As soon as the seat was clear of the aircraft, a drogue parachute would deploy stabilizing the seat in a free fall. When the seat fell below ten thousand feet, a larger parachute, triggered by a pressure switch would open and slow down the seat and occupant.

If the larger parachute were to open at a high altitude, the occupant may not survive the extreme cold temperatures and his bail-out oxygen would not last during a long slow descent. At any time below ten thousand feet, the seat occupant was required to undo the seat straps, holding him to the seat, kick away from the seat, pull his personal parachute ripcord, and gently float to the ground to a safe and happy landing.

No one even discussed grounding the aircraft to fix the knee contact problem. Design and structural changes would take a while. Believing that my small stature would prevent me from requiring a wheel chair in the unlikely event that an ejection was to occur, we set off to Boscombe Down.

Extensive War games were taking place the morning we departed and we were told to expect mock encounters with RAF fighter aircraft.

We were given the radio frequency they would be using. To avoid an encounter, we climbed above 65,000 feet. It was amusing to hear the cry tally-ho as we were spotted by squadron of fighter aircraft and the frustration and surprise they expressed when they could not climb high enough to get near us. It made us realize that our spy plane was indeed a winner.

B (Bomber) Squadron of the Aircraft and Armament Experimental Establishment had been given the third production Canberra PR 9 for acceptance evaluation. On this aircraft, they conducted handling and performance as well as navigation and photo camera assessment.

On arrival at Boscombe Down, we were provided with instrumentation to measure the receiving and transmitting signal strength of several antennas located on the aircraft.

Tests were conducted at different heights and at multitudes of headings, covering a full three hundred and sixty-degree angular range.

There was a long, flat beach area on the coast where it was recommended that we conduct land/sea crossings to evaluate the difference in the radio altimeter signal as it rebounded from water to land and visa versa. After several runs in opposite directions, a clear difference of height when over land or water was established. I vaguely remember that the difference was less than ten feet. On our way out to the beach, I had noticed a hill on the outskirts of the Town of Yeovil, in Somerset.

I thought it would be a good idea to measure the height of that hill and compare it to a map reading. Dickey Turley George complied with my request and flew straight toward that hill. As we came to the top of the hill I was alternating between reading the radio altimeter and looking through my new periscope sight when suddenly, there was a strange loud noise, and through my periscope, I saw the biggest helicopter ever built at that time, accelerating up toward us. It was a very near miss and again I thank the Gods.

It turns out that on the other side of that hill was the Fairy Rotodyne Test Center. The vertical-lift aircraft that they were developing was years ahead of its time. It was designed to carry up to seventy-five passengers at very high speeds. The rotors were powered by high velocity air bled from two Napier turbo prop engines. It was reputed to be extremely noisy, which we could both agree was correct.

Shortly after we landed at Boscombe Down, our pilot was reprimanded for flying in a restricted area and was informed that the Fairy Rotodyne had temporally lost control due to the wake turbulence it encountered as we flew over. I felt pretty bad not only realizing that it was my request to fly over that hill but also that I could have caused a serious set back in important research work to say nothing of an awful serious accident and potential loss of lives.

Before I end this saga of awkward incidences on the PR 9, I wish to convey the genuine esteem I held for that magnificent and robust aircraft.

It seemed to have a personality of it's own, while yet being an inanimate but eye pleasing object. I would often pat the nose of the aircraft after returning from a test mission, as I did with my horse Velvet, many years previously in a different world at a different time. There were times during the strapping-in procedure before, flight that I wondered if I would ever make it back to unstrap. Whether I was in an ejector seat or not, bailing out was going to be a hazard. With the ejector seat, I ran the risk of being decapitated at the knees. Without an ejector seat, I stood the risk of getting into a tight body spin if a bail out occurred at high altitude. A safety bulletin had been issued by the RAF and spelt out the hazards of high altitude bail out. Crews were warned not to deploy their 'chutes at altitude, due to the extreme cold temperatures and insufficient emergency oxygen pressure to maintain breathing and suit pressure. During free fall prior to 'chute opening, a high velocity body spin could result in brain hemorrhage, causing death. Crew, were advised to move arms and legs in and out to vary the body center of gravity to prevent the dangerous spin. Autopsies carried out on dead airmen who had landed when their parachutes automatically deployed at ten thousand feet, bringing them gently to the ground, revealed evidence of death by brain hemorrhage. If I were ever unlucky to bail out without an ejector seat, at high altitude, I would have remembered the Hokey Kokey, which for those of us old enough to remember, was an energetic group dance. It goes like this:

"Put your left arm in, put your left arm out, put your left arm in and shake it all about, you do the Hokey Kokey and turn around about, that's what it is all about. Ooooh the Hokey Kokey, Ooooh the Hokey Kokey. Knees bend, arms stretch ra ra ra." Yes, I would do the equivalent of The Hokey Okey all the way down until my parachute opened.

All the flying on the PR 9 was completed and I was busy polishing up the final touches to my report, when the newly appointed director of flight operations patted me on the shoulder, apparently for a job well done and promised to buy me a beer when the aircraft was accepted by The RAF. My first meeting with my new boss was many years earlier during my first employment with Shorts. He was the pilot who went berserk when I had worn his May West.

The aircraft was soon accepted and I never did get my beer. The PR 9 test program gave me a rough ride to Certification, the roughest ride I ever encountered before or after. My reward was the pride I felt in my contribution towards producing a great aircraft.

The communicating antenna equipment aboard, supplied by the United States, allowed the transmission of photographs in real time to intelligence centers anywhere in the world. The technology used was the same as that used on the Lockheed U 2 Spy plane. The PR 9's flew deep into the Soviet Union and China prior to and later in co-operation with the American U 2 program.

The PR 9 and the U 2 aircraft were both capable of operating out of reach of enemy fighter aircraft. Note: the maximum height Alex and I reached during tests on the PR 9 aircraft was Sixty-seven thousand three hundred and seventy feet. An altitude not much lower than the famous U 2 spy plane shot down by a ground to air missile over Russia, with Gary Powers aboard. At sixty seven-thousand feet, the air pressure is ¾ psi (pounds per square inch). This contrasts with 14.7 psi, which is the air pressure at sea level. Air pressure in outer space is 0 psi, which is not too different to ¾ psi. Thus, the high altitudes that were required to fly were near equivalent to being in outer space, except that gravity is present.

At such altitudes the familiar sky-blue colour changes to a sky as seen just passed sundown. The dark and somewhat purple shade gave me an out-of-this-world sensation. At heights around sixty-four thousand feet, I could see that the earth was indeed round, as the horizon clearly shows the curvature of the earth. Any flat earth society member would resign if he could be up there with me.

The twenty-three aircraft built remained in service with the RAF for well-nigh half a century. The last official flight took place to honour the Queen's 80th birthday, when the aircraft flew over Buckingham Palace surrounded by an escort of fighter aircraft. While I was involved in the Canberra and other research projects, major changes were taking place in the company. A contract from the ministry of Supply was obtained to produce thirty large military freighter aircraft to be known as the Belfast. The aircraft could carry two hundred fully-armed paratroopers and was also capable of carrying tanks and heavy armored vehicles.

CHAPTER 13
Farewell to Shorts

    A large hanger had been built to house several of the very large planes to be used on the test program. Prior to the arrival of the first aircraft, a vast team of pilots and test engineers occupied the swish office space in the new building. On my first visit to the new facilities, I came across several new-fangled coffee machines. I placed the appropriate coin into a slot, selected coffee with milk and awaited the paper cup to drop down to be filled with a steaming hot drink. The hot coffee came out all right but with not a cup in sight, drenching my shoes, socks and trousers. I bought my own cup for use next time.

    The next surprise was the posting of the new organization structure, which noted me as a Section Chief responsible for Air Conditioning and Fuel Systems. Having been responsible for many aircraft test programs, covering all subjects, I was disappointed and somewhat insulted as many new employees were given more senior positions. Later on the next day, I was called into the director's (prima donna's) office and as though being bestowed a great honour, I was ceremoniously granted a key to an "executive" washroom. I was told not to reveal this treasured gift to my colleagues. I was so taken aback that I left the room with the key, without saying anything. It soon dawned on me that having an executive washroom carried the class system too far. The eating areas in the company was another example of the English class distinction, practiced in Ireland. The junior weekly-paid staff and all factory workers used the main cafeteria. The senior weekly-paid staff was fed in a large room with tables holding four people, with the same choice of food as that in the main cafeteria. Monthly-paid staff, were fed in a room similar to the senior weekly staff, the food was slightly better, and was served by waitresses. Junior executive staff were fed similar to the monthly staff but had table cloths and napkins on the tables. Senior Executive staff, were fed in a room with a much greater ambience and, at lunch, were offered complementary beer or wine with their meals. The Board dining room was out of bounds so I cannot describe it.

The audacity of having such a thing as an executive washroom in a closely-knit Flight Test Organization got to me and I made an appointment with the Managing Director of the Company to register my objections.

A few days after my meeting, the lock was removed from that "special" washroom.

Having completed all the test plans for air conditioning and fuel systems on the new aircraft, I did not wait for the aircraft to arrive and decided it was time to move on.

The English Electric Aircraft Company offered me a position as a senior flight test engineer on their new and futuristic fighter aircraft, to be known as the TSR 2 (Tactical Strike and Reconnaissance Mach 2).

The TSR 2 conducted its first flight in September 1964, a year after I had been offered a position to fly in that aircraft. An interesting episode, relayed to me later was that during early test flying, both crew-members were blinded at a critical time during landing. When the undercarriage was selected down, vibration, associated with hydraulics, matched the frequency of the human eyeball in its socket, resulting in blinding the crew during the critical landing phase. On April 1965, the program was cancelled due to complications and cost overruns.

As luck would have it, I had decided not to join English Electric but to go to Canada and, join the de Havilland Aircraft Company. It was a move I did not regret.

During eight years of living in Ireland, apart from enjoying the work, I enjoyed the country and its people. The "troubles" had not yet restarted and yet there were many tell-tail signs of trouble to come. Apart from the segregation of Protestant and Catholic by streets and areas, within the Aircraft Company, there was religious segregation among the shop workers. This segregation was not obvious to me until it was pointed out. For example, sheet metal workers and electric installers were all members of one religion and machinists and mechanics belonged to the other religion. The unemployment rate of Catholics was significantly higher than that of the Protestants. Marching bands would parade through "opposition" districts inciting riots by playing their nationalistic or rebel tunes. Northern Ireland was the only place I knew that banned singing in Pubs to prevent fights from breaking out.

There was a house that had been vacated by a Catholic family and over several nights, we could hear the tribal beating sound of drums, which kept us awake for most of the night. After enquiring about that noise, I was told that a protestant family was due to move into that vacant home and those sounds we heard were coming from Lambeg drums, which were traditional protestant marching drums. They were beaten to drive Catholic devils out of the house.

Living in Northern Ireland was never dull. The glens of Antrim and the seaside resorts on the beautiful coast road, were a joy to behold.

Swimming in the Irish Sea was not for the faint hearted, it required a combination of grit and foolhardiness to plunge into that cold but invigorating body of water.

For several years, we lived in a caravan (trailer home) and since there were no trailer parks at that time, I ran around trying to find a place to park. A sheep farmer whose farm was on the outskirts of the small village of Carnmoney, offered to let me stay in his sheep pen which contained a sheep dip. We were told that sheep would only be herded in there for short periods about every four months. The sheep dip was right in front of the living room window and gave us hours of entertainment watching sheep guided into that chemical dip. Some would get into the dip without any problem and others would put up fierce resistance, giving the farmer and his sons a hard time. The sheep were herded into our "front yard" the night before their dip. They bleated all night and occasionally one would get its horns stuck in the brake cables, which were beneath the floor. The constant bleating and occasional banging was not as bad as the day they completed the dip and left our front yard covered in sheep droppings. It wasn't the only time that droppings were strewn around the front doorstep. On one occasion, I was late for work and had not yet emptied the chemical toilet, which was embarrassingly full to the brim. I pushed the overfull toilet, carefully trying not to spill the contents and rested the heavy load on the edge of the doorstep. As I was preparing to jump over it, the load tipped and the contents were deposited all over the place. Having avoided the long hike to the midden, where I was supposed to empty the contents. The toilet was now empty, I set off to work leaving that awful mess behind. I quite understood that my wife was not too pleased.

I owned a 1935 Austin Seven which was a box-like car with steering that required full and constant attention. Steering was completely unstable. If you took your eyes off the road even briefly, the car would dart towards a hedge or ditch. It was time for a better car, so I traded it in for a 1936 Rover, which had plush upholstered leather seats. Proud as punch, I drove it home from the dealer. Before I got home, I decided to check the reverse gear. It worked fine, but the gear would not return to forward. I drove the car home backwards, parked it and used it as a home for my children's pet rabbits. I purchased a large 1936 Chrysler, which was completely out of place in the narrow Ulster country lanes. On one occasion, the Chrysler's engine stalled just after we had left the driveway of the farm. Out I came with the heavy cranking handle. It was hard to turn the crank and one had to be cautious of kick-back which could break an arm. Exhausted, after many attempts at starting the engine, I pulled the crank out of the engine and with a display of temper, I slammed the heavy iron crank into the pavement, it bounced up and knocked an elderly man off his bicycle. I apologized and helped him to his feet and straightened out the twisted handlebars of his bicycle. He helped my wife and kids push-start the car and off we went to the seaside.

When it came time to bid farewell to Ireland, we did so with many lasting memories and looked forward to a new life in Canada.

Adventure in the Air — 105

DH Buffalo Military Transport

DH Buffalo with load extraction 'chute deployed

DH Buffalo Cargo drop incident

DH Buffalo ACLS (Air Cushion Landing System)

DH Buffalo Augmentor Wing Research A/c

Augmentor Wing with Author (left), Don Wittley & Seth Grossmith

Tracker Fire Bomber

DH Twin Otter in Yellowknife

DH Twin Otter Water Bomber

DH Twin Otter Skiplane

DH Twin Otter Floatplane Water Bomber

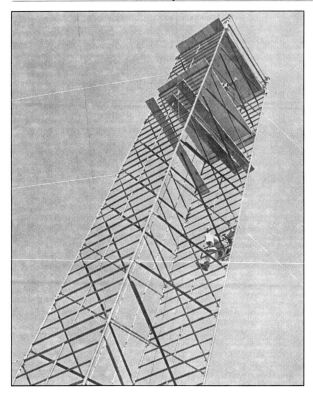

Camera tower in Flagstaff Arizona

DH Dash 7 Transport A/c

Adventure in the Air — 111

DH Dash 8 Transport A/c. Water Ingestion Test

DH Dash 8 high angle abuse take off

CHAPTER 14

De Havilland Aircraft Company of Canada (1963-1991)

De Havilland Canada was formed as branch of the English Company in 1928, the year I was born. At the time of my arrival, Phil Garratt was Managing Director of the company. He was a hand on manager, knew his employees well, and was very approachable and well liked. He set the tone for a well-run company.

A British Overseas Aircraft Corporation Constellation aircraft flew me from London to Toronto. My wife and family shared time with her parents and my mother until I found accommodation in Toronto a month after arrival. We now had three daughters, the youngest being only one year old.

A colleague from Shorts, Jock Aitken and his wife Jean met me at Toronto's airport. Jock had immigrated to Canada a few month's previous. While staying with them, I found a home to rent in Bolton, just north of Toronto. A month after my arrival in Canada, my family joined me.

The Flight Test Department at Downsview, Toronto was similar to the departments at AV Roe and Shorts during my early days there. Large Panoramic windows overlooked the airfield. The manager of the department, Wally Gibson was also a new employee, and my initial impressions of him being a fine boss to work for, were never dashed. Two pilots were also located in the department, Bob Fowler was the Chief Experimental Pilot and Mick Saunders was his deputy. The Instrumentation and Data Reduction Departments were adjacently located.

My initial assignment was to be a senior test engineer in charge of the Buffalo test program. It was three months prior to the first flight of the Buffalo. Instrumentation was being installed, as well as flight test safety features. These included a rocket at the rear end of the fuselage, to provide nose down pitching in case a deep stall was to be encountered. Explosive bolts were fitted to the rear entrance door for quick emergency exit. Knotted ropes were installed leading to all possible exits, to provide handholds in case of abnormal aircraft attitudes.

Sharp edges were rounded off and padding applied to protect the test engineer from cuts and bruises.

It was the modus operandi at de Havilland for the senior test engineer to constantly monitor and provide test direction for all test flights by radio communication with the aircraft. He had specialists at his beck and call. It was a good and sensible approach to testing, not unlike my previous experience with single seat aircraft. I reached an agreement with management that it was important that I maintained some flying experience and could occupy the jump seat between the pilot and copilot on flights I would be interested in participating in.

This was a busy time. The Turbo Beaver was conducting certification tests. Four Buffalo aircraft were conducting test programs. An Augmenter wing Buffalo research program was taking place in California. The Hydrofoil high-speed Canadian Navy ship, called the Bra d'or, was about to start sea trials in Halifax Nova Scotia. The roll out of the Twin Otter was immanent. Yes it was a hectic time. Wally Gibson showed his strength as an organizer and we often burnt the midnight oil, scheduling and planning tests.

Experienced flight test engineers were hard to find. We managed to place on contract, two test engineers from Lockheed, in the USA. Stuart Nicholson joined us from Shorts. Barry Hubbard and Bob Dingle were the company stalworths. Jock Aitken traded his flying suit, used on the Turbo Beaver, for bell-bottoms, as he accompanied the Canadian Navy on the FHE 400 hydrofoil. The HMCS Bras d'or was conducting high-speed sea trials and achieved 63 knots. The high speed runs were listed as flights as the foils lifted the hull completely above the surface of the sea.

We hired a handful of keen, but inexperienced engineers and watched them show promise, as time went on.

During my early days at de Havilland, many interesting tests came my way. The first of which was a Beaver aircraft modified to conduct aerial spraying of locust swarms. This was especially interesting to me, as many years previously, I had witnessed locust gathering in the outskirts of Peshawar in Pakistan. Baffles were installed to prevent locusts clogging engine cylinder cooling fins. An oil cooler bypass was also installed for similar reasons. A spray bar and chemical tank completed the modifications. Engine and oil cooling tests as well as aircraft handling evaluation with the long external spray bar soon cleared the modifications. Attack of the locusts could now begin.

Another modification of the Beaver aircraft resulted from a request from the US Army. The Beaver was to lower a Black Box attached to a cable. This Box was to be winched down on a one thousand-foot electric cable to dangle well below the aircraft. In a very tight turn, it was required to ascertain the height of the aircraft that would allow the Box to hover one hundred feet above ground. It was an interesting task and even more interesting to surmise the nature of the Black Box. It was during the Vietnam war and rumor had it that the Box was a sniffer capable of smelling the scent of a human hidden in a jungle, where the Viet Cong would be hidden.

The US Army gave us another interesting task. This was to install a furl able antenna device on the rear ramp door of a Caribou aircraft. This antenna was produced and provided by Spar Aerospace, a Canadian Company devoted to Space research. A wide flat coil of spring steel, when uncoiled would roll itself into a rigid tube. It was somewhat akin to a large spring-loaded tape measure. At the end of the tube was a large hook. We were to lower this tube below the aircraft and report on its stability in the air stream at various air speeds and then wind in the tube. The tube was remarkably stable in the air stream and could be extended and retracted. We were led to believe that the device was to be used to hook onto a parachute and haul in the occupant, to prevent him from landing in enemy territory.

De Havilland Buffalo (1964-1967)

The DHC-5 Buffalo was a military tactical transport aircraft designed to have STOL (Short Take Off and Landing) capability. It was the successor of the DHC-4 Caribou and had a larger fuselage capable of carrying a 3/4 ton truck or a 105-mm howitzer, and was powered by two General Electric T-64 turboprop engines, replacing the Caribou piston engines.

In the spring of 1964, the aircraft commenced high speed taxi tests. A lot of high speed runs and short skips off the runway were accomplished, thanks to the low speed flying qualities of the aircraft. So many runs were made that the brakes got too hot and were binding, preventing first flight. The brakes were taking a long time to cool and the flight might have been delayed due to impending loss of daylight. Being impatient, the company Chief Engineer, Fred Buller, ordered the ground crew to pack the wheels continuously with snow, to accelerate cooling. A successful first flight took place with all the usual glowing comments.

In an attempt to expand initial speed restrictions, the aircraft encountered a severe elevator control flutter, accompanied with significant vibration, and speed was immediately reduced. After advising the crew to stand by, I contacted a senior engineer to acquaint him with the problem and reluctantly relayed his advice to Bob Fowler, who was the pilot in charge. The advice was to go back to the speed where flutter was encountered, place one of the 25 lb ballast bags on the control column and check the effect, and add another bag if necessary. After a long pregnant pause, Bob asked me if I thought it was a wise move, I said, I did not think so. Meanwhile, Bob Dingle, the test engineer on board, had painstakingly untied the safety netting and extracted some ballast bags. With the ballast bags handy, Bob Fowler decided to go ahead. With one, then two 25 lb bags on the control column, speed was increased ten knots above the original flutter speed, without encountering any flutter. That was enough experimentation and the flight was aborted. The information gathered enabled the elevator balance weight to be increased and thus eliminated the flutter problem.

When stall tests commenced, I was on board the test aircraft and sometimes on board on the chase plane, accompanied by aerodynamic engineers. Having experienced stalls on many aircraft, I was surprised at the heavy buffeting and loud banging noise, especially with full landing flap. The banging was due to the twisting of the rear fuselage, interacting with the large rear closed a ramp door. During the banging, daylight could be seen between the ramp door and fuselage. Tightening the door by adjusting the door actuator shut off switches, improved the problem. Later when watching the aircraft stall from the chase plane, the fin T tail junction rocking movement was alarming to watch and no doubt caused the rear fuselage twisting. A post-stall-flight special inspection was instigated to ensure fin/tailplane attachments and structure remained intact after each stall flight. With approximately a hundred stall flights conducted, the structure remained intact. Note: an aircraft should not encounter a stall in its lifetime. Stalls are now practiced for crew training in simulators, if available.

De Havilland Aircraft of Canada was justifiably renowned for producing aircraft that could take off and land from short fields. The Buffalo was no exception.

Prior to the introduction of the Buffalo, a research program was completed, which investigated steep approach and landing in order to define limitations due to pilot capability.

The research aircraft was a highly modified standard single engine Otter aircraft with a jet engine installed in the fuselage behind the pilot and test engineer. The jet engine had nozzles pointing slightly forward and down, and could be vectored by the pilot to provide mainly negative thrust with plenty of drag. Two Pratt and Whitney turboprop engines were mounted on the wings, replacing the standard nose-mounted piston engine. A special landing gear was installed capable of withstanding high potential impacts. There would be plenty of stretch in the extended tube to reduce the shock on potential heavy landings impacts.

Bob Fowler, who achieved approach angles as high as nineteen degrees, flew the research test program. The nose-down attitude of the aircraft was significantly greater than nineteen degrees. This required the aircraft to be rotated approximately twenty-five degrees before making the correct touch-down attitude. I can imagine the fear factor encountered as the runway appears to move rapidly up towards the aircraft and the pilot rotates early to prevent a heavy impact. This was the absolute limit, which an average pilot would never encounter. Resulting from Bob's work, a seven and a half degree approach angle and a fifteen hundred feet a minute descent rate was considered a steep approach limit.

A low operating speed for take off and landing is an essential ingredient for operating out of short airfields. A 1500 foot field length was the target for short runways. Major runways are 10,000 ft long. Another essential ingredient was the capability for a steep climb-out and as already noted, steep landing angles. Conventional aircraft land at a three degree angle. The Buffalo, when in a military tactical role, was capable of landing at a steep angle of seven and a half degrees. The advantages of steep angles for take off and landing, is that any obstacles in the vicinity of airfields, such as buildings, trees, hills, towers and so forth, can be safely cleared. Touch-down scatter during landing can also be reduced. Another advantage is that a steep spiral take off and approach can be made in the confines of a secured runway area. This would provide protection from enemy small arms fire. The United States Army and Marines showed great interest in the Buffalo program.

There were no civil regulations at the time, permitting reduced take off and landing distances by reducing speeds or conducting steep approaches. For military operation, there were no ground rules either. To plan the tests for tactical military take off and landing tests, I used my own guideline:- "Keep it short and keep it safe". With the advantage of having Bob Fowler's office next to mine, I was able to consult with the leading expert on STOL operations, and other company experts. The end result was a 15 % reduction in conventional take off and landing speeds using maximum take off and landing flaps. An engine failure during take off was not to be considered. It was accepted that the aircraft could not survive a statistically rare engine failure during take off. Maximum reverse power on both engines was to be used immediately on touchdown. A valuable aid to the flight crew was the Fast/Slow Indicator. The indicator received its inputs from the stall warning system. When the indicator needle was centered, speed was correct at 15% above the stall. If the needle went to the left in the red zone, speed was too slow. At the end of the red zone, a stick shaker would be triggered, warning of an impending stall. To the right, the needle went into the green zone indicating that speed could be reduced.

Sitting in the jump seat between Bob Fowler and Mick Saunders during steep landing tests, was to say the least, invigorating, like a roller coaster ride.

Measured test results provided data that demonstrated STOL operations could be undertaken from such short runways that the aircraft could take part in a demonstration in New York. The exercise was to support a simulated emergency evacuation of the city. The Buffalo flew into a ballpark in Manhattan. Here, a mobile hospital was unloaded. Meanwhile, a Twin Otter landed on a nine hundred foot long boat peer on the Hudson river, to demonstrate emergency evacuation from the heart of the city. Was this in anticipation of 9-11? Some time later a Buffalo landed in a car park in downtown Tucson Arizona during an US Marine supplier's convention.

City center operation was a goal which took a long time to achieve, due mainly to noise objections from nearby residents. A noise reduction program took place later, allowing Dash 7 and Dash 8 de Havilland aircraft to operate without noise pollution, from the London Docks, the Belfast Harbor airport and the Island airport adjacent to the heart of Toronto.

Previous experience had taught me that operating away from base, with a small team, paid great dividends. Conducting tests where weather statistics were in your favor, runways were long and air traffic controllers provided traffic free areas for tests, ensured maximum utilization of time spent in the air. The psychology of being given the responsibility without interference from headquarters was great for team spirit and moral.

My first remote test site location for de Havilland was in Marana Arizona. Here, a small team, completed in four weeks, tests on a Twin Otter that would have taken approximately six months to complete at the Toronto base. Such results set the stage for many future remote operations. They were all very successful, allowing me to present a paper on Remote Site Testing to the Society of Flight Test Engineers in Seattle and to the Canadian Aeronautics and Space Institute in Ottawa.

Arizona provided not just good weather but also hot weather in the summer for hot weather tests. There were nearby airfields where high airfield tests could be conducted. Most of the performance testing on the Buffalo, Twin Otter, Dash 7 and Dash 8 was conducted away from base.

The Buffalo performed many aircraft performance and hot day tests in Arizona. Tests were for both military and civil operation. The aircraft was developed with weight increasing from 38,000, 43,000, 46,000 pounds. Performance tests were conducted at 49,000 pounds, but the aircraft was taken out of production before this weight was certified. Each of these weights required extensive testing.

It happened to be a mild winter day back in Downsview when the opportunity arose to investigate operation under wet slush conditions. The snow was turning into slush, so we got the airfield bulldozers to pile slush on the runway. The side exit doors on the Buffalo were removed to allow me to obtain video pictures of any slush entering engine intakes or any other vulnerable areas. I had installed a safety line to prevent me from falling out of the side door. Runs through the slush were conducted at several speeds and configurations. During the runs, I leant far out through the open doorway to get good video coverage while relying on the safety line to prevent me from falling out of the aircraft. When the tests were successfully completed, I attempted to remove the safety line and to my horror the attachment hook had somehow come undone. I had been leaning precariously out of the doorway without a care in the world. Perhaps the Gods held me back, preventing a tragic fall.

One of the away from base operation took place in Peru. The Peruvian Air Force had purchased eight Buffalo aircraft and were due to receive a further eight. The aircraft were to be used, in some instances, to operate from new small airfields located along the Amazon and on the East of the country. To get to these small jungle strips, the aircraft was required to fly over the Andes Mountains. Two problems were encountered that prevented flights over the Andes. These problems justifiably caused a hold in payment of the first batch of eight aircraft and the refusal to accept the delivery of the second batch.

The problems were that the engine oil pressures would fluctuate violently above 25,000 ft and the fuel control unit did not supply the required power at altitudes above 25,000 ft. These problems had to be rectified in a hurry. The aircraft had previously been tested at 25,000 ft without displaying any of these problems.

It was soon realized that there was not enough pressure in the engine oil tank at high altitudes to prevent cavitation on the suction side of the oil pumps. This caused air to be sucked into the oil system causing fluctuations. To investigate a solution, our brilliant and practical Chief Designer, Fred Buller, came up with one of his typical ideas. We took on board, a wheel and tire pumped up to 30 pounds per square inch and connected a hose from the tire to both oil tanks. At 34.000 ft. there was no sign of oil pressure fluctuations. A pressure regulator enabled me to lower the pressure until fluctuations of oil pressure occurred. This allowed us to select and design a pressure supply and provide a permanent solution that would eliminate the problem.

A "product improvement" had been introduced by the engine manufactures to the fuel control unit. We were unaware that this so called improvement had not been verified at high altitude. A simple fix to the electronics of the fuel system was incorporated and required testing.

The Buffalo is an un-pressurized aircraft and we conducted many flights at altitudes around 36,000 feet to ensure the modifications were satisfactory. On one occasion I was sitting in the jump seat and felt a significant pain in my right shoulder. The aircraft had an amazing rate of climb and in fact obtained a world climb record in its class, demonstrating a time of eight minutes and three and a half seconds to reach Twenty-nine thousand five hundred feet from the start of take off. I knew I had got the bends and should have been breathing oxygen a while before take off, to purge nitrogen from my blood stream. I continued to complete the logging of engine data before informing Bob Fowler of my problem. He conducted a rapid decent and immediately the pain disappeared.

On the following day which was a Saturday, I woke up and noticed that my right shoulder was swollen to about twice it's normal size. It was painful, but not unbearable. My wife called the doctor and surprisingly, he paid a house call. After examining my shoulder, I was advised not to worry, take a couple of painkillers and call back in a few days if the swelling did not go away. By Monday morning the swelling had increased further. I went to work and showed off my bulbous shoulder. At no time did I connect my malady to getting the bends a few days previously. As soon as Bob Fowler noticed my swelling, he called the Institute of Aviation-Medicine, which was located across the airfield. They immediately placed me in a pressure chamber and I spent eight hours at high pressure representing many feet below sea level. I was later informed that if I had delayed much longer I would not have survived. I was indebted to the Gods once again.

A team of product support personnel was dispatched to Peru and installed the fixes on all eight of their existing aircraft. The Peruvian Air Force insisted that each aircraft had to be tested with one of their pilots as a witness. The aircraft were to operate on all the new routes across the Andes Mountains. Bob Fowler and myself were chosen for this interesting mission. The sales department wanted to impress the Peruvians that they would be sending their best team, so my title as Senior Test Engineer was changed to Chief Flight Test Engineer. It was strange way to get a promotion. Later I became Manager of the Experimental Test Department. My task was to record data and issue reports to the Air Force to satisfy them that the aircraft performed to specification and in particular, had sufficient climb power at high altitude to match the aircraft flight manual.

We arrived in Lima and before long were ready for our first demonstration. The aircraft was loaded to a heavy weight, which was a good thing as the rate of climb to altitude would be slower than that encountered when I had got the bends back home. Accompanying us on board was the Commander of the Transport Squadron and an Air Force Flight Mechanic. There were only three oxygen regulators aboard which would allow pressure breathing. We advised the Squadron commander that we could not accommodate the Flight Mechanic but he insisted, saying that his Mechanic could use the passengers' oxygen tubes. Off we went, I was recording climb and engine data while sitting in the jump seat between the pilots. Standing behind me was the Flight Engineer with an oxygen tube stuck in his mouth. As we reached 30,000 ft, I noticed that he was now sucking on two oxygen tubes. By the time we got to 34,000 ft he was showing obvious signs of discomfort, so we rapidly returned, left him behind and continued the flight and completed the data gathering.

The following day, we set off across the Andes with another Air Force pilot named Esquardo. It was agreed that there was no need to test above 30,000 feet. Esquardo pointed out our destination. It looked extremely tiny especially from high altitude. I remember thinking that landing would be well-nigh impossible. We came in lower, and saw that the strip was not much longer than a soccer field. Esquardo suggested that we conducted a low run over the field to check it out. He was in the right hand seat and Bob handed him the controls. After lowering the gear, he flew low, very low over that field. He was so low that the wheels brushed the tops of the jungle trees surrounding the strip. Bob took back the controls and we landed with not much room to spare. As a matter of fact, we were unable to turn around and had to go into reverse to back up. I was the first to exit the aircraft and was startled to see crowds of Machete wielding natives rushing toward the aircraft. As they came close to me, they stood to attention, saluted and shook my hand. A lot of hand shaking occurred as the rest of the crew left the aircraft. The headman of the nearby village was a relative of Esquardo and the two were genuinely pleased to see each other. There were monkeys in the trees as well as parrots and some colourful birds and large butterflies. The air was full of oxygen and a delight to breathe in. It was like paradise and reminded me of my days in an Indian jungle.

Soon however, we took off gazing down on some majestic peaks of the Andes and returned to our base in Lima, looking forward to the next adventure.

Our next trip was a flight to Cusco. En route we recorded climb performance and engine data. We circled over the famous ruins at Machu Picchu. It was hidden away in the mountains and not surprising that this jewel was not discovered for a long time.

After arriving in Cusco, we were asked to deliver some corrugated roofing and some army personnel to a small strip not far from the city. The troops arrived and loaded the roofing material. They were all small in stature, not one of them reached beyond five feet high. My five foot five inch stature allowed me, for once in my life, to tower above so many others. I sure felt like a big man. We arrived at a small strip, deposited the troops and roofing material and delayed departure to Lima until a number of pigs were herded away from the runway.

The next assignment was to a penal colony. It was located across the Andes in a dense jungle area. Once again data was recorded prior to arrival at a luscious green field, which was soaking wet from a heavy rainfall the day before. Our wheels gouged deep tracks on the runway.

We were met by the Commander of the colony and taken to lunch beside a fast flowing stream. We were told that there were no cells or fences; any prisoners wishing to escape could do so. There was nowhere to go and survival in the jungle was pretty well impossible. Prisoners were brought here by air, hands bound and legs shackled. On arrival, they were free to find food and shelter in an area we did not see. The dining area was open sided, covered by a thatched roof. The air was good to breathe and there were plenty of birds and butterflies to gaze at. Lunch consisted of a large bowl of green soup filled with rice, accompanied by chicken and sweet potatoes. Topping the soup was a slice of green cucumber-like vegetable. I noted that the soup surrounding this vegetable was extremely hot and spicy, so I set it aside. Before I had time to warn him, Bob Fowler picked up his slice and placed it in his mouth. Immediately thereafter, the hot spice flavor was very evident and he rapidly removed it with his fingers. His eyes started to water and rubbing them with his contaminated fingers made matters much worse. After the laughter died down, we realized that Bob was in a lot of discomfort. He was taken to the stream to wash his eyes, which remained red and painful for several hours, delaying our departure. We were all glad to leave as none of us wished to stay overnight with the "bad guys" in a penal colony.

After arriving back in Lima I was able to spend a few days analyzing climb and engine power data and preparing the format for presentation to be given to an Air Force General after all eight aircraft had been tested. In the evenings, after dining, Bob and I would explore the many city squares within walking distance of our hotel. On one evening, we went to a movie to pass the time away. The movie was a Hollywood comedy, dubbed in Spanish with no subtitles. The laughter and enthusiastic responses to a lot of the scenes were a joy to behold.

Never did either of us witness such enjoyment emanating from the big screen.

CHAPTER 15

While dining one evening, there was a sudden rush for the exit door; before we realized it, we were in an earthquake. There was severe rattling of crockery but not much else. The restaurant was soon empty of all its clients, except for Bob and I, who had not the same survival instincts as those who had experienced many earthquake disasters. A severe earthquake had devastated some remote areas in the country a month before our arrival in Peru. The Buffalo aircraft were engaged in round-the-clock missions conducting earthquake relief.

The next aircraft we were given to test had just come back from transporting sick and wounded earthquake victims. The aircraft had to be hosed down to clear the aftermath of many upset stomachs. The pilot and flight mechanic who were to fly with us, showed obvious signs of fatigue, especially the mechanic who looked extremely tired and in need of a long rest. During our flight to Iquitos, I had just completed my climb and engine data gathering. We were at 29,000 feet, still over the mountains when I turned around from the jump seat and noticed the Mechanic lying flat-out on the side canvas seats. He appeared fast asleep and his oxygen tube, which should have been in his mouth, could be seen dangling away from his head. We were not yet able to descend due to the terrain. I removed my oxygen mask went back to assist him. At first he did not respond and I feared the worst. He did stir however after some violent shaking and I managed to place the plastic end of the oxygen tube into his mouth. Being unable to locate a second oxygen tube, I was obliged to share the oxygen with him until we were able to descend. We took turns sucking the tube end. Because of his obvious state, his share of the oxygen was greater than mine. Soon we were able to descend and he recovered with no signs of ill effect. This demonstrated very clearly, the negative safety connotations of fatigue. We landed in a small field beside a small town, and delivered the sacks of rice and flower we were carrying. It was late afternoon and the Peruvian pilot was anxious to get back. Meanwhile, the flight mechanic uncovered the missing oxygen tubes, which were in a plastic bag and adding the one tube we shared, he took them to a crew room nearby to be disinfected.

Soon after we got airborne there was an agitated conversation in Spanish. When translated by the pilot, it turned out that the oxygen tubes were left behind. We were due to fly back over the mountains at 29,000 feet with only three oxygen supplies, and there were four of us.

At first, the Peruvian pilot refused to return and told his mechanic to suck on the oxygen supply line that the tubes were connected to. This was a ridiculous suggestion because, not only were the oxygen supply lines high up near the top of the fuselage side wall, but the tubes to be used were to be plugged into non-return valves, so oxygen could not be sucked out. Much to his chagrin, he was talked into landing to pick up those missing tubes. We got back to Lima at dusk without any further incident.

Two more successful flights across the Andes were conducted to Attillia and Masa Marie.

At a meeting with the squadron commander, we were told that we were only required to fly one more mission to a high altitude airfield.

This would mean that we would test seven of the eight Buffalo aircraft.

After each flight I had handed over a report containing tests conducted and recorded data. Analysis of data and comparison with the flight manual was to come later in a final report.

The high airfield we were to go to was, Juliaca, which was over 13,000 feet high. It was a challenging airfield for many aircraft as witnessed by several wrecked aircraft in the vicinity of the airport We had the squadron commander aboard again and once again, I recorded the data during a climb to 30,000 feet. After landing, we were asked to park the aircraft for several hours to allow the engines to cool down. A cold engine would be more difficult to start at this high altitude.

Moreover, the auxiliary power unit (APU) used to provide bleed air to start the engines had never been cleared to start at this high altitude. If the engines could not be started, we would be spending some time in Juliaca. Waiting for engines to cool, gave us time to explore the town. From the airport we could see lake Titicaca on the border between Peru and Bolivia. We set off on foot to explore. I had forgotten my camera and ran back about 400 feet to the aircraft and retrieve my camera. I was huffing and puffing by the time I caught up with the others. As I got to them, Bob noticed that my lips were blue, due to lack of oxygen at this altitude. Looking at him, I noted that he too had blue lips. There was also another incident that pointed to our high elevation. We were passing by a high brick wall about six feet high and frightened a chicken. It attempted to fly up to the top of the wall. After a lot of wing flapping and a long struggle, its feet reached the top of the wall but, it's center of gravity was below the top. It flapped it's wings madly trying to gain lift. Slowly and surely the bird reached its goal and rested at the top of the wall. With wings spread and drooping down, mouth wide open gasping for air, that bird gave a prefect demonstration of the difficulty of high altitude flight.

We walked through a market, had a meal in a small restaurant and sampled fish from lake Titicaca.

Three hours later we were back at the aircraft wondering if failure to start would give the Air Force an excuse to cancel the contract. The first attempt at starting the auxiliary power unit failed. It however, did start on the second attempt and provided the bleed air to start the main engines. Both engines started without a problem and soon we taxied out to line up for take off. A mountain ridge was straight ahead and most aircraft would circle to gain height before attempting to fly over that ridge. We impressed the Commander by flying straight over the ridge with plenty of height to spare.

On return to Lima, I worked hard in my hotel room to prepare the report. The elimination of oil pressure fluctuations was a "slam dunk."

No fluctuations were witnessed on all twelve engines tested. The engine climb power on all twelve engines were shown to be in excess of the engine manufacturer's guarantee. Measured climb performance was compared to the aircraft flight manual and shown to be superior, due to the excess of power available during the tests on the six aircraft. When engine power was corrected down to the guarantee level, it was shown that the flight manual was correct.

The General insisted that he only wished to see Bob Fowler and myself. He did not want company executives or salesmen at the meeting.

The Squadron Commander was also present. The General thanked us for our efforts. My report was handed over and a summary page was placed in an overhead projector. It took less than five minutes to discuss the table and chart.

We stopped for coffee and sandwiches, exchanged pleasantries and the meeting was over.

Our experience in Peru was very rewarding. Not only did we go to places that most tourists would never see, we were also able to operate the aircraft in an operational environment and into and out of very short fields. It is one thing to conduct short take offs and landings out of a small section of long runways, it is entirely another thing to take off and land from postage-stamp size runways in a remote jungle strip that, at first glance, looks impossible to operate out of.

Shortly after our return home, all the remaining eight aircraft were delivered to Peru. Mission accomplished.

Cargo dropping.

The Buffalo rear cargo loading design made the aircraft an ideal candidate for aerial delivery. Two types of delivery were to be tested.

HAD (heavy air drop) and LAPES (low altitude precision equipment supply) I knew that initial cargo dropping tests were hazardous and since no one had cargo drop experience, and we hired a consultant to give us advice.

He was from the US Air Force and gained his experience on Lockheed C-130 aircraft. The one day spent with him left us in no doubt of the hazards, after he described some catastrophic failures due to lack of attention to details and, errors associated with cargo and parachute rigging.

The following day we attempted to conduct the first airdrop of 4000 lbs. The drop zone was to be at a military base near Welland, Ontario.

The technique for a heavy airdrop is as follows:

Rollers are installed on the floor of the aircraft to create low friction. After carefully rigging the large drop-parachute on top of the drop pallet, the smaller extraction 'chute is attached to both the palette and to the rip cord of the large drop 'chute. Just before the drop, the small extraction 'chute is installed in a bomb release clip located above the wide-open rear area. The extraction 'chute is released at approximately 1000 ft above the drop target area. Immediately on release, the 'chute inflates and pulls the load out. As soon as the load exits the aircraft the large drop chute inflates and then lowers the load gently to the drop area.

During our first drop attempt when the extraction 'chute was released, nothing happened except for a slight movement of the extraction pack in the bomb release clip. It was a bad start. Here we were, with a heavy load poised to release at any moment. We had passed the drop area and Bob Fowler, who was flying the mission, flew low avoiding any built up areas, in case an extraction inadvertently occurred. I was tempted to go to the back and secure the hesitant extraction package but prudence prevailed. I did not want to become part of the load that may be suddenly extracted. After declaring an emergency, Bob carefully landed. We wisely decided to find a safer and more remote area to continue the cargo drop exercise.

Marana in Arizona was the logical choice. Here we had a convenient isolated drop zone, and a specialized parachute club who had worked with the military to pack all sizes of parachutes. We were also given permission to use an area beside a disused runway for low-level cargo drops.

The first two heavy air drops with 4000 lb and 6000 lb went well. We had learned that a simple breakaway lanyard attached to the extraction parachute would prevent the previous predicament we encountered at Welland.

The 8000 lb drop was a near disaster. Unknown to us, the company supplying the aluminum drop palettes on which the loads were secured, had replaced the drop palette with a transport palette, which was significantly lighter and less costly. When extraction commenced, the palette lugs broke away, leaving the load behind. The extraction 'chute then proceeded to extract the large main heavy airdrop 'chute from it's position on top of the palette. The large main 'chute deployed and pulled the load with great force and being rigged for a vertical drop, was attempting to rotate the palette inside the aircraft. All this happened in a split second. As the palette left the aircraft in almost a vertical position, some damage to the rear fuselage occurred. We were lucky the whole tail section of the aircraft was not torn away. The Gods were once again with us.

I believe the representative who supplied the offending palette felt pretty badly. He sold the damaged palette for scrap and, with the one a thousand dollars received, we held a barbeque and invited a large number of employees from the base.

After repairing the damage to the rear fuselage, we completed the heavy airdrops by successfully dropping a 12,000 lb load.

Low altitude precision extraction supply (LAPES) drops were next on the agenda. The technique for this type of drop was to extract a load while flying just a few feet above the ground. This was done by deploying an extraction 'chute, sometimes multiple 'chutes to pull the load out from the open rear of the aircraft. The undercarriage is selected down in case of inadvertent contact with the ground. The extraction 'chute is deployed early and is attached to the floor at the rear end of the fuselage on a special clevis. On demand, when a few feet above ground, the clevis would transfer the extraction force from the floor of the fuselage to the load to be extracted, thus causing rapid extraction.

The tests were completed, when three palettes each containing 4000 lbs were linked together and dropped successfully. It was rewarding to see how accurate these precision drops were. On one occasion, with Barry Hubbard at the controls, the undercarriage was festooned with branches from shrubs highlighting that these tests were indeed carried out at very low altitudes.

The cargo drop program was also used to familiarize several pilots with cargo drop techniques.

Armed with my newly-acquired experience, I was dispatched with pilot Bill Loverseed and a sales team to Honduras to demonstrate the newly acquired cargo drop capability of the Buffalo aircraft.

We arrived in Tegucigalpa in Honduras and awaited the call to conduct a demonstration. United States Military Advisers were staying at the same hotel and, on occasions, we were kindly given rides to the airport in their bus. During four days of inactivity, we enjoyed their company and often dined out together.

On the fifth day, the aircraft was loaded to conduct a 8000 lb Low Altitude Precision Equipment Supply (LAPES) drop. Prior to the drop, I accompanied a Colonel of a Honduras Parachute Brigade. We were flown in a small military helicopter to the proposed drop zone in a jungle area. The Colonel was a picture postcard character. He was short, thick set and sported a large wax pointed mustache. He had on knee-high riding boots and jodhpurs and carried a swagger stick with which he would often strike his boot with a resounding crack. He spoke English well, so communication was not a problem. We arrived at a small grass airstrip about 1000 ft long, located in a valley with a small hill at one end and a meandering stream at the other end. It had been raining heavily the previous night and the stream was swollen and flowing fast. There were a large number of paratroopers gathered around the area and I was informed that they would be performing a para drop after the LAPES demonstration. I informed the Colonel that there were no troop seats installed. He brushed off my statement saying lack of seats were no matter.

The aircraft arrived within radio range, and I was able to communicate with the crew, describing the low hill over which we had to descend. Because of the hill, the crew was advised that the drop was to commence half way down the strip and just beside it. They were also advised to land after the drop and we would discuss the feasibility of dropping paratroopers.

The drop occurred exactly half way up the strip and the load went on and on and on, ploughing in and out of the stream near the end of the runway and ending at about a hundred feet beyond. The load traveled an unusually long distance after contact with the ground due to the extremely low friction of the wet grass beside the airstrip. When we arrived at the load, I saw a number of small fish, wriggling and gasping for air on the load palette. I picked up a handful and deposited them back to the stream, much to the amusement of some of the paratroopers. They did however help me to return some of the stragglers back to the stream.

After the aircraft landed, we agreed to drop the paratroopers. There was no man count; they all piled in and sat on the floor in several rows, with legs spread apart grasping the floor rollers. The take off was conducted with the cargo door open and the ramp flush with the floor. During a steep climb after take off, those troops must have gripped the floor rollers for dear life, as they might have fallen out during the acceleration and climb out. The drop occurred from approximately 1500 ft above the ground. They came out like clockwork until two of them collided and their chutes entangled with each other. There were cries from below shouting "emergencia, emergencia" to no avail as the two were engaged with words and fists, battling each other on the way down. Eventually the emergency 'chutes were pulled open just in time to land safely. After landing they continued their battle only to be separated by their comrades. It appears that one man hesitated, resulting in their collision thus causing the fracas.

After we got back from this exercise, the Canadian Embassy asked us if we could transport some equipment to a Canadian missionary hospital in a remote area not accessible by road. We of course agreed, and picked up the equipment, which had been in a field near the airport at San Pedro Sula. The equipment had been awaiting transport for many months and the missionaries were desperate to get it. The equipment consisted of a tractor and a crated x-ray machine. As we began to load the crate holding the x-ray, several iguanas scurried out of their home in the crate. On our way to the mission hospital, the aircraft was crawling with ants, presumably displaced from their home in the crate.

We landed on a small grass strip at the mission used only for small light aircraft, to beaming smiles from priests and nuns. There were cheers as we unloaded the long awaited tractor and x-ray equipment. A quick lunch and we took on board a dozen people destined to return to the capital. After arriving back and exiting the aircraft a young soldier suddenly swung the butt of his rifle to my thigh and knocked away a large black hairy tarantula, he then stepped on it leaving a gooey mess. We did have a creepy-crawly flight.

There was one more task for the aircraft. We were asked to display the aircraft at the opening for a new military air base. The President of Honduras was there and amidst great pomp and ceremony, we watched a colourfully adorned colour guard goose-stepping along the parade ground. The band struck up to play the national anthem and then proceeded to play Danny Boy. I could not help but smile as it seemed strange to me to hear this Irish tune played at a Honduran base opening ceremony. It was entertaining but somewhat out of place.

The next day we headed home. A week later, we heard that five US Military advisers were killed by a bomb exploding in their bus as they left the hotel we all stayed at. Once again the Gods were with us, as we had often traveled in that bus with them.

Some time later, a stop-and-go cargo drop was proposed. The technique we used was simply to conduct a short landing in the drop zone, stop and then immediately accelerate to take off and, by doing so, drop the load out of the rear open cargo area. It was considered to be a low cost operation without the use of parachutes.

We took a Buffalo aircraft to Mountainview near Trenton Ontario, installed a 4000 lb load on a standard transport palette. Aware of potential hazards, I decided to modify the technique. After landing and coming to a complete stop, we increased power to the take off setting with the wheel brakes on. We then released the load hold-down and the wheel brakes, allowing us to accelerate rapidly and drop the load at the start of the take off run. The load dropped about four feet onto the runway just a couple of feet away from where it was positioned in the aircraft. We increased the load in increments to a maximum of 10,000 lbs with similar results. Satisfied that the technique was safe and simple, we demonstrated the stop and go drop to the company sales engineer. He suggested that the four foot drop may not be acceptable and a that the Americans conducted gentler drops from the C 130 Lockheed aircraft by lowering the rear ramp door closer to the ground. We had set the ramp door to be flush with the cargo floor and pointed out that lowering the ramp door would make no difference as the high longitudinal acceleration would propel the load beyond the ramp and still land from a height of four feet. We were told that he had seen the Americans accelerate slowly, dropping the load gently to the ground. Very reluctantly I agreed to investigate his suggested technique.

We loaded the palette to 4000 pounds. Due to a wind change were obliged to use a shorter runway. After touching down, there was a slight delay while the load was unlocked. When acceleration commenced, the aircraft was less than one thousand feet from the end of the runway. The load moved reluctantly toward the rear, and did not appear to be in any hurry to exit. We now had two concerns. The slow aft moving load would cause the airplane center of gravity to exceed it's aft limit, causing unstable flight. Abandoning the take off would require heavy braking which would cause the load to come crashing towards the crew area. the crew compartment. At the last moment, Bill Loverseed pulled the nose up, we dropped the load near the parameter of the airfield and climbed safely away.

Needless to say no further investigation of this technique took place. Once again, the Gods need to be thanked.

Transporter.

A civil version of the Buffalo military transport was developed and was named Transporter. The aircraft had some changes to its structure, which allowed an increase in take off weight. When the weight increase trials were completed in Arizona, we were asked by our sales department to find a typical desert scene to locate the aircraft for still photographs.

We met with the team in Marana and learned that they had scouted out a location on the border of Arizona and California. There, we were to meet with two Arab students, a brand new Land Rover with a driver and a person titled the coordinator, and a couple of camels that hailed from a zoo. The Photographic team that joined us aboard the Transporter came from a Hollywood studio they consisted of a photographer, his assistant, and a wardrobe mistress. The photograph to be taken was to represent the delivery by air of a Land Rover, to two Arab Sheiks.

The wardrobe had been placed in a protective covering and was carefully placed on the floor on top of ballast boxes. We took off mid morning.

En route, leaving my jump seat position, I wandered back to converse with the photo team, and sat down on the ballast box on top of the wardrobe. Waving hands, a disapproving look and a frantic shriek, alerted me that I had done something wrong. I leaped up and was told not to wrinkle those precious robes. That ended any potential conversation and I went back to my seat in the cockpit as the wardrobe lady was smoothing out imaginary wrinkles.

The map coordinates led us to our destination. It could have been in Arabia as it was a vast expanse of desert with nothing in sight. It was a rough landing, due as we found out later, that gopher holes were all over the place. It was a credit to the rough field capability of the undercarriage.

It took several hours to find the perfect location of the photographer relative to the position of the aircraft and where the sun would set. The photographer was elderly, portly and gruff. He treated his young assistant as a lackey. The coordinator arrived at about 3 pm with the Land Rover and the two Arab students.

There were no camels. After harsh words between the coordinator and the photographer, the show went on without the camels. When it became time for the students to get dressed, they refused to don the robes provided by the wardrobe lady because they were peasant robes and could not be warn by sheiks. As luck would have it, they brought their own robes. I could not help but smile at the wardrobe lady's embarrassment.

The two pilots and two sheiks posed patiently in front of the Land Rover and Transporter, awaiting that magic moment when the sun was just right on the horizon. It was most entertaining watching the photographer and his assistant work their trade. After struggling to get down in the prone position, at the bewitching moment the photographer barked out orders. He kept his assistant busy changing lenses and filters, cleaning them and wiping the sweat off the man in the prone position. The sun soon set, the task was completed and a great advertising picture later graced auto and aircraft magazines. I accompanied the aircraft later to the Farnborough air show in the UK, taking the opportunity to spend a bit of time with some of my relations there. On board the ferry flight, was a reporter from a Canadian Aircraft Magazine. Halfway across the Atlantic Ocean, a loud bang was heard coming from the avionics console accompanied by a small trace of smoke and a slight smell of burning wires. To those of us experienced with test and development aircraft, it was nothing much to be alarmed about. Pulling a couple of circuit breakers isolated the problem, which would be fixed on the ground later. We had just lost one of the many navigation aids, which was not essential. The reporter however was extremely nervous and maintained a frightened look until he was calmed by noting the absence of concern from the rest of the crew on board.

Experience does bring calm to tense situations.

CHAPTER 16

While Tom Appleton and Barry Hubbard demonstrated the aircraft during the air show, I met up with several old colleagues, including Jack Hines and Alex Roberts of Shorts. Alex had become Vice President of Sales for Shorts and later Tom Appleton became Vice President of Sales for de Havilland.

After a few days after visiting with friends and family, I rejoined the aircraft on its return to Canada via Iceland and Greenland where demonstrations were to take place for potential customers.

It was in Iceland that I got my first glimpse of the Northern Lights, also known as the Aurora Borealis. I had left my room to join the rest of the crew in a restaurant across a field, when I saw the spectacular night sky in an unbelievable display of dancing lights and colours of green, red and silver, shimmering and making curtain calls across a dark sky. I was not expecting it but there it was, Rembrandt could not have painted a better picture. I was truly awestruck and could not bring myself to leave that live theatrical show curtesy of the heavens above. I have seen the Northern Lights, since this first encounter; they pale in comparison. Needless to say, I was late for dinner.

When we got to Greenland, we were tasked by a Norwegian-owned aircraft carrier, to take the aircraft to a remote island which had been used as an emergency landing strip during World War 2. It was a short and narrow gravel strip, desolate and uninhabited.

We awaited the company evaluation team to take them to the remote strip. It was to be a two hour flight to get there and only two and a half hours of daylight left, when they arrived, complete with wives and children. We took off in a hurry in order to get there and leave before dark, as there were no lights to guide us. We got there; it was narrow and during the turn around for take off, the nose wheels got embedded in the gravel, right up to the wheel axles. It was getting dark and we were well and truly stuck. We stood around gawking, until Barry Hubbard came out of the aircraft with the "honey bucket" (portable toilet), he empties it and proceeds to dig out the nose wheels and form a ramp.

With lots of power the aircraft climbed out of the hole with the accompaniment of cheers, all directed towards Barry. We got back safely and no one was marooned, cold and hungry on that remote island.

ACLS Buffalo C8A Research Aircraft.

Instead of a conventional landing gear, the Air Cushion Landing System (ACLS) has an inflatable but perforated rubber and nylon air cushion bag, which permits operation from and to almost any type of surface. The Bag is inflated by bleed air from fuselage side mounted Canadian Pratt and Whitney engines, like a hovercraft, the aircraft can operate from ice, water, snow, rough fields, soft soils and swamps. When deflated, the bag hugs the bottom of the aircraft without causing aerodynamic drag. Bell Textron developed and manufactured the bag.

Initial flight assessments to check out modifications and inflation and deflation of the air bag, revealed the expected difficulty in low speed ground handling. Directional control, especially under cross wind conditions was to say the least, tricky. The four brake pads, which protruded below the bag, were not very effective. The aircraft, like a wind vane, was constantly heading into wind. At low ground speeds, the brakes and rudder provided little control. Lack of control during the start of a take off run, can be described like being in a giant hockey puck being pushed around the ice by sticks of wind. This feeling suddenly ceased when speed built up to approximately fifty miles an hour. Independent use of the left and right throttle levers provided some contribution to directional control on the ground. An improvement in low power engine thrust response by combining propeller control with engine control had been contemplated for a Buffalo product improvement program. This improvement known as beta control would certainly enhance the poor directional control noted during the initial evaluation of the air cushion landing system. Beta Control was rapidly incorporated and benifitted both the ACLS Buffalo and the basic Buffalo aircraft. The aircraft was flown to the Wright Patterson Military air base in Ohio. Here it underwent further evaluation during a joint program with representation from the US Army and Bell Textron. Here at Wright Patterson, we were introduced to the strict protocol for operation at that base. We were allocated a low priority and were 23rd on the priority list. Even to go outside the hangar and run the engines we were required to obtain a permit, which required twenty signatures. Permission to fly required even more signatures, and a formal briefing with base representatives. In spite of the strict protocols and low priority, we received tons of cooperation and the tests proceeded at a reasonable pace.

Handling and performance tests were undertaken without incident until the aircraft was undergoing stall tests. The aircraft was not fitted with any stall recovery device. Previous tests on a standard Buffalo indicated that such a device was not necessary.

When a stall was conducted with the air cushion bag inflated, a deep stall was encountered. The aircraft developed a very high rate of sink, rapidly heading for the ground. There was virtually no response when the pilot pushed the stick forward in an attempt to increase speed. All three controls, elevator, ailerons and rudder were unable to cause any aircraft response, and initially were essentially useless. The stall had commenced at a "safe" height around twelve thousand feet. Normally with the Buffalo only a few hundred feet were typically lost in recovering from a stall. After about a couple of thousand foot height loss, our pilot, Bob Fowler, who was the company's Chief Experimental Test Pilot, and I may say the best test pilot I had ever flown with, was about to tell us to bail out, when ever so slightly, he noticed the nose just barely beginning to drop. Shortly thereafter sharp buffeting was felt on the elevator controls, indicating that airflow had reattached at the horizontal tail and elevators. The controls became effective and the aircraft was finally recovered with some altitude to spare. I was required to thank the Gods once again.

Tests at Wright Patterson were soon completed and the aircraft was delivered to the Canadian Forces base in Cold Lake Alberta, for operational evaluation on ice, snow and water. Tests were completed by the Canadian armed forces and the program was shelved due to lack of funding.

Augmentor Wing Buffalo Research Aircraft.

If the basic Buffalo aircraft can rightly be described as a world leading STOL (short take off and landing) aircraft, the Augmentor Wing Buffalo research aircraft can only be described as a Super STOL aircraft.

The Augmentor Wing is the result of a co-operative program between the US and the Canadian Government. The concept began in 1960 at AV Roe in Canada. After the ill-fated cancellation of the Arrow program, Don C Whittley, who was the Project Leader of the Augmentor Wing Research Group, joined de Havilland to continue his pet project.

The Canadian Department of Industry Trade and Commerce and NASA, funded a program which combined de Havilland, Boeing and Rolls Royce Canada, to build and test a research vehicle using a highly modified Buffalo aircraft. When this program was completed in 1980, the aircraft was returned to de Havilland.

The augmentor wing concept can be described as follows:-

High velocity compressed air from a jet engine is directed to split flaps at the rear of the wing. The right engine provides air to the left wing and the left to the right wing. The temperature of the air is not high as it has not yet been subjected to burning by jet fuel. The high velocity air exits the rear of the wing through a specially shaped slot, which causes surrounding outside air to join it, thus augmenting the air flow. This sheet of air when directed backward, provides forward thrust and when directed downward, provides some vertical lift. A major bonus is obtained by inducing the surrounding outside air, to join the sheet of air exiting the rear of the wing. The bonus gained is that at very low air speeds the air flow over the rear portion of the wing is maintained thus delaying flow brake down and reducing stalling speed. This allowed very slow speed operation.

Forward thrust and vertical lift are also provided by the hot jet exhaust which, could be also directed back or down by means of rotatable jet exhaust nozzles..

The augmentor wing concept allows low speed operation without the attendant high drag penalty.

For the technically-minded reader, the augmentor wing is described as follows.

High wing lift is generated by a combination of blown flaps and vectored thrust.

Cold by-pass air from the two jet engine compressors, is cross-ducted to a spanwise, specially shaped slot in the rear of the wing just in front of the wing trailing edge flaps. Due to the shape of the slot the Coander effect causes ejector action to accelerate ambient air from the rear top surface of the wing through the flaps and thereby increases thrust and reduces wing drag by creating boundary layer control. The left engine supplies air to the right side ducts and the right engine to the left ducts. When flaps are lowered, the sheet of high flow air, points downwards and contributes to 35% increase in vertical lift. This is augmented by improved airflow over the wing, adds additional lift, allowing low speed operation. Downward jet thrust from exhaust nozzles, which can rotate, provide the vectored thrust also providing additional 65% to vertical lift.

Early in 1981, Don Whittley came into my office and told me he had obtained funding for a last ditch effort to impress the Canadian government and open the door for many applications of the Augmentor Wing concept. Don gave me complete freedom to propose an appropriate flight test program, and wished me to direct that program. We were still in the midst of certification tests on the Dash 8 aircraft and poorly placed to add another task for our Flight Test Department. I was torn between my responsibilities for existing programs and my keen interest in the development of short take off and landing capability. The augmentor wing Buffalo, represented the pinnacle of slow speed flight and I would have give my eye teeth to get involved. My boss, Wally Gibson was away on holiday in Florida. I called to acquaint him with the dilemma and informed him that the program could be conducted in Mountainview, which was less than a two-hour drive from our plant in Downsview. The holiday spirit was in him, as bless his soul, he gave me the OK, and I was able to embark on a program that had the potential of placing Canada in the forefront of an aviation breakthrough. Apart from Don Whittley's design group and Bob Fowler, the aircraft was relatively unknown to the mainstream of the Company. A learning curve had to be undertaken by the small team assigned to test and maintain the aircraft. Bob Fowler could not be sprung loose to pilot the aircraft. We obtained the loan of Seth Grossmith, a pilot from Transport Canada. He had been seconded to NASA and had undertaken many flights on the aircraft. I had previously flown with Seth several years earlier, during certification flights on a Grumman Tracker water bomber.

In 1965, a large-scale model of the concept was tested in the wind tunnel at NASA's Ames Research Center in California. Successful completion of the wind tunnel program led to the building of an aircraft to demonstrate a proof-of-concept. The prototype Buffalo was modified significantly by a shared contract given to de Havilland and Boeing aircraft.

Major modifications of the original Buffalo aircraft included a significant reduction of the wing span by 17 feet and replacement of all of the wing structure aft of the rear spar. The air distribution ducts were installed as well as the new flaps, which could direct augmented airflow, back or down. The turboprop engines were replaced by Rolls Royce Spey 801 jet engines, with rotating nozzles obtained from the vertical take off Harrier aircraft.

It was not until May 1, 1972 that the aircraft conducted its first flight from the Boeing base in Seattle with Thomas Edmonds in control. The aircraft was flown to the Ames Research Center in California, where many pilots were invited to fly it, including Bob Fowler.

After evaluation of the low speed characteristics afforded by the augmentor wing, the program at NASA changed direction and concentrated on area navigation and automatic terminal guidance. Joint government research contracts ended in 1980 and the aircraft was returned to Canada. The National Research Council in Ottawa, who had designed the automatic flight stabilizer system, completed modifications to the control system. Short evaluation flights of the auto stabilizer system were carried out by Stan Kerulek who had flown the aircraft when it was operating under the control of NASA. The aircraft was flown to de Havilland and parked beside our test hangar. It was then that Don Whittley gave me the opportunity to get involved with the program.

The test instrumentation on board the aircraft had been installed by the National Aeronautical Establishment in Ottawa and could only be played back at the de Havilland Company's mainframe computer with the inherent delays due to being off site and other high priority work. A simple quick-look instrument package was installed to measure take off and landing performance.

Previous cameras methods which were used, had mechanical problems and with no spare parts available, the cameras had been disposed of. The trisponder was unable to be integrated with the NASA instrumentation on board the aircraft. The procurement budget would not allow purchase of expensive equipment. Necessity being the mother of invention, an inexpensive solution was found. The trisponder, a radar distance measuring device, was placed in a vehicle to be located at the end of the runway and its receiver was located in the aircraft.

Visiting a popular electronics store, I came across an infrared burglar alarm, which would sound off as soon as an intruder interrupted a beam. The alarm was purchased and placed in my car as I drove by my home, where my wife held the reflector in the doorway. Every time I drove past, pointing the alarm at ninety degrees to the small reflector, the alarm would sound. The working range was only about one hundred feet. The next day the burglar alarm was taken to work and Don Band, who was also an electronic wiz, easily found a way to boost the infrared signal. Now the alarm would operate at a three hundred foot range. A two-foot wide and four-foot length of half-inch plywood, covered with reflective material, provided a large reflecting surface. The total cost of the alarm and reflective sheet on plywood was approximately seventy-five dollars. The desk-top computer used for analysis cost about five hundred dollars.

The alarm was mounted facing out of a side window in the aircraft and was wired to trigger an event on the instrumentation in the aircraft as soon as it passed by the reflector. The large reflector was placed at the edge of the runway in a surveyed location.

An old dodge car owned by Don Band was placed near the end of the runway. The computer sitting in the passenger seat of the car received digital signals from the required parameters in the aircraft. These signals were sent by the VHF audio link normally used for voice communications. Trisponder data was also gathered in the vehicle's computer.

The start and stop point aircraft position was noted. Don was instrumental in putting this instrumentation package together. In this way, every take-off, landing or aborted take-off always had at least two runway positions were identified, allowing correlation of data recorded on the aircraft with data recorded on the computer in Don's car. Subsequent analysis (on a desktop computer), provided a calculated distance. Analyzed data obtained after each test run could be examined before the next test commenced.

The test program was written to demonstrate the aircraft's capability of meeting existing certification standards as well as proposed standards for powered-lift aircraft. Approach and landing techniques, and maximum assault take offs, were to be investigated and recorded. Since the aircraft, in a later test phase, was due to demonstrate operation from an aircraft carrier, a navy landing technique was also included in the program.

Much to my delight, Seth Grossmith invited me to occupy the copilots seat for initial flights, after which he was due to train de Havilland pilots Bill Loverseed and Barry Hubbard to fly the aircraft.

The 20th August 1982 was the date of the ferry flight to Canadian Forces Base in Mountainview. Wanting to make every flight count, I had included an initial in-flight assessment of a very low landing speed. During the ferry flight at a safe height, we simulated a landing and set up the chosen low speed. At this speed, we conducted turns and pull-ups in increasing increments until we reached half a "g" without encountering any problems. This demonstrated that the proposed special rules for landing speeds were met.

Once in sight of the runway at our destination, we noticed many heavy vehicles on the runway and were informed that the main runway was closed. We were unaware that a NOTAM (Notice To Air Men) had been issued regarding the closure of the runway. A small cross-runway used mainly for small aircraft, was open for use. For our initial arrival, we had planned a conventional landing, which would have meant landing at nearby Trenton. Having just completed a simulated short and steep landing, we were confident that the short runway was adequate and it gave us the opportunity of starting the program right away. Sitting in the cockpit during a steep and slow landing was like descending in a fast elevator, with the ground coming up fast to meet us. A firm touch down and a rapid deceleration caused me to be thrust forward against my seat straps. In a remarkably short ground roll, we came to a halt in a few short seconds. If we were competing for the best amusement park ride, we would had been sure winners. We landed with plenty of runway to spare, much to the amazement of the adjacent runway workers.

Over the course of my career I had witnessed over one thousand engine cuts during take offs. The negative effect on directional control was always evident. Not so with the Augmentor wing. When for example, the Port engine fails during take off, the augmentor airflow, from the Starboard side, will still maintain approximately half the thrust. The Starboard side, will at the same time, loose approximately half the thrust, as it loses the augmentor flow from the Port engine. Thus an engine failure on one side, will result in reduced but almost equal thrust on both sides.

Handling qualities at all the low speeds investigated, met and surpassed certification standards.

Another pleasant surprise, was in the serviceability and reliability experienced, underlying the simplicity of the augmentor wing concept.

Stunning performance of the aircraft carrier landing simulation was recorded. Glide path control was excellent causing very little scatter in the target touch down point, which was only plus or minus ten feet. The straight-in landing without flair, as used by navy aircraft, produced an average touchdown sink rate no different to any conventional landing technique. Landing ground roll distance, when corrected to a 30-knot headwind, which is guaranteed to be available on the deck of a carrier steaming into wind, was less than 200 feet without the aid of an arrester hook.

The assault take off technique demonstrated gave equally amazing results. The technique used was based on all engines operating. The engine exhaust nozzles were rotated just before lift off to thirty-six degrees down. This provided additional lift while still maintaining some forward thrust. In zero wind conditions, the take off ground roll was only three hundred and thirty five feet. At an equivalent weight and wind conditions, the take off ground roll of a standard Buffalo aircraft is 60% greater. When converted to a thirty-knot wind, which represents a minimum wind on an aircraft carrier, the ground roll would be only eighty-six feet. These results indicate this aircraft which is bigger than most carrier aircraft can take off without the aid of a steam catapult and land without the aid of an arrester hook. This could be accomplished in one-fifth of a typical 1,000 foot long aircraft carrier deck.

During the test program, a team from Lockheed Aircraft visited the test site. They were interested in the Augmentor wing technology and sent an evaluation team, including a pilot to fly and witness some of the tests. They were amazed at the stunning short field performance of the aircraft. Having just acquired a million dollar laser tracking system to record and analyze the same type of airfield tests we were conducting, they were equally amazed at the economical and simple measuring methods we were using.

The de Havilland Aircraft Company of Canada at this time was up for sale.

You would think that Lockheed would have bought our Company, having been so impressed. But no, it was Boeing that bought the Company. And that is another story.

In addition to the then current test program, the augmentor wing design team had been working on a special application of the concept which provided vertical lift to a futuristic fighter aircraft equipped with a very fat, short span and low-drag augmentor wing. Although the wing was small, the depth of the wing could hold an enormous amount of fuel. Vertical lift was achieved by relative cold augmented air flowing from the center fuselage area and thus avoided the hot exhaust re-ingestion problem of previous vertical take off and landing aircraft. A large scale wind tunnel model had undergone wind tunnel testing with encouraging results.

Armed with such spectacular results, a meeting to request further funding was held in Ottawa with the Canadian Industry of Trade and

Commerce and the Department of Defense. A comment at the end my report is as follows :-

"Development of the thick transonic wing and the quiet three-stream jet engine, when coupled with the outstanding low speed qualities demonstrated by this research aircraft, will provide a quantum leap into the STOL technology for which Canada is respected. When caution is weighed against opportunity, let us hope that the multi-purpose potential of a quiet jet STOL aircraft as demonstrated in this and future test programs, will not be lost through the economical restraints present today."

Funding was not made available; navy personnel would never witness the spectacular take off and landing performance on a carrier. Further development ceased; the augmentor wing research team was disbanded.

Many Canadians are aware of the demise of the Avro Arrow. Unlike the Arrow, which in it's day dominated fighter aircraft design, the augmentor wing project is relatively unknown and deserves equal credit, which alas, will not be bestowed. Aircraft, for the last thirty years have suffered from lack of innovation as product improvement on existing aircraft rules the day. The obscenely high cost of developing new types of aircraft today prevents such products from ever taking shape. In my early days in aviation, I had witnessed many innovations introduced by a handful of dedicated personnel at minimal cost. Those hay days of aviation are long gone and I wonder why? The automobile industry has only recently introduced new concepts, like the hybrid and soon, I hope, will introduce hydrogen powered and fuel cell electric powered cars. Don Whittley's augmentor wing concept could have achieved a quantum leap in aircraft design. It was not to be ,and now the augmentor wing concept is displayed in the Canadian Aviation Museum in Ottawa, a sad reminder of opportunity lost.

CHAPTER 17
DH Twin Otter

On May 20 1965, without much fanfare, the prototype Twin Otter arrived in our crowded test hangar and joined the four Buffalo test aircraft. A short while later the Twin Otter test fleet grew to four aircraft. At the same time the Hydrofoil ship, the Bras d'Or was conducting sea trials near Halifax. To describe our test operation as hectic, during those days and years, is an understatement. It was crises management most of the time but somehow we managed to juggle problem solving, install fixes, plan safe test procedures and participate in test flights. I had always believed in the team approach and teams were assigned to each aircraft. It was rewarding to note the rivalry between teams as they compared flight test hours achieved on each of their projects.

Prior to its first flight, the prototype Twin Otter conducted highspeed taxi tests on the main runway at Downsview. Being a short take off and landing aircraft, the "taxi" runs enabled the aircraft to get airborne to a height of about two hundred feet. This was accomplished in various configurations, providing confidence in the handling behavior of the aircraft. When the official first flight took place the following day, it felt like an anti-climax. The Twin Otter followed in the footsteps of the Standard Beaver and Otter. It was robust, simple and reliable. A production batch of eight hundred and forty-four aircraft indicated the popularity of this aircraft. The aircraft had a simple fixed undercarriage and had two Canadian Pratt and Whitney PT 6 Turboprops mounted on each wing, providing increased safety over the standard single piston-engine Otter. It could carry twenty passengers. This reliable Canadian aircraft is still operating in both urban and remote areas spanning the globe.

The first Twin Otter was assigned to investigate handling qualities and aircraft performance. After being bogged down resolving unacceptable stalling characteristics, the performance test phase commenced.

When the time came to record engine-cut take off data, a serious situation occurred. The aircraft was being flown with Mick Saunders as pilot and Bob Dingle in the co-pilots seat as test engineer. Maximum power was set for take off and brakes were then released, causing the aircraft to accelerated down the runway. While still on the runway and just before Bob was due to use his cut switch to rapidly shut down the left engine, Mick noted that power on the left engine had crept up above the maximum red line and throttled back to reduce power. This occurred just as the test cut speed had been reached and Bob cut the engine. Normally, a couple of seconds after the cut, the propeller on the failed engine would feather, thus reducing propeller drag. Mick had inadvertently throttled the left engine too far back just prior to the cut.

This action disarmed the auto-feather system, causing a lot of drag and prevented the aircraft achieving a climb capability. While Mick was busy controlling the aircraft, Bob was equally busy attempting to restart the engine. The aircraft had barely scraped over the perimeter fence and was flying dangerously low over a fuel tank depot, which was just beyond the fence. They were flying in-between two large fuel container tanks and were below the top of the tanks when power was restored and the aircraft was able to climb to a safe height and return to base. The gods were with them that day.

I was monitoring by radio, this flight and another test flight on a Buffalo aircraft. The incident caused me to propose that we cease all potentially hazardous testing from Downsview. The surrounding area was far too built up. A near catastrophic disaster had been avoided by the heroic efforts of the crew.

After studying weather statistics, it was determined that Arizona would be a good location. We were just passed the critical stage of development when close support from all the engineering disciplines was constantly required and most foreseeable modifications had been incorporated.

It was no longer essential to be based at home. The location we chose was Marana Arizona, which was just outside Tucson. Here, unhampered by built-up areas except for cattle and a stockyard, we started remote site operations. This heralded other remote locations where we tested many future aircraft projects. Remote site testing paid great dividends and later resulted in a paper published in the Canadian Aeronautics and Space Journal. The introduction in my paper could be of interest to the reader and is presented below:-

"It is becoming increasingly difficult to launch a new aircraft program and meet current airworthiness standards. The good old exciting days when companies launched new aircraft projects every year or every few years are gone. Research aircraft are becoming museum specimens, and a ten year cycle between new projects is now an expectation, based on development costs.

At de Havilland, we have managed to keep flight costs on our test projects low. This has been achieved by minimizing expenditure on data acquisition systems and by finding the right remote sites to enhance cost effective testing. A two- and-a-half million dollars saving and a reduction of seventeen aircraft test months can be directly attributed to the success of remote-site testing.

Canadians migrate south at evert opportunity to get a brake from the harsh winter. The weekend pilgrimage to cottage country is also a Canadian tradition. Remote-site testing is therefore a natural extension of these migratory habits.

The flight test fraternity of all companies has one thing in common- dedication and enthusiasm, this being the result of the nature of the task, i.e. the romance of flying and the challenge of the unknown, with some thrills and travel as added bonuses. Dedication and enthusiasm are somewhat constrained by the wide range of influence of home base operation and domestic obligations.

Short-term, remote site testing at the right location and with the right equipment for independent operation, provides the incentive for accomplishing goals. This approach is shown to be cost effective".

Marana Arizona not only provided good weather, it permitted hot weather testing in the summer months and nearby were several high altitudes to permit high altitude airfield performance testing. There were very few air traffic constraints, and we could conduct tests without being continually told to change direction or altitude. We had also planned to conduct high altitude airfield tests in Flagstaff, which had an airfield elevation of just over 7000 feet. The one runway here was in a shallow valley surrounded by many tall pine trees. Our performance camera and survey theodolite had to be placed above the trees to obtain a view of the aircraft during take off and landing. A local contractor had been tasked to build a tower for the camera.

We arrived in Marana, to conduct aircraft performance testing, with a work crew totaling seven. A pilot, test engineer, instrument technician, mechanic, inspector, photographer and myself, who led this, the first of many successful remote test operations. The prototype Twin Otter was crammed with spares, mainly consisting of brakes and tires, test equipment including a portable wind station, the large performance camera and personnel baggage. Eager for an early start the next morning, we surveyed a location for our performance test camera, filled ballast sandbags with readably available desert sand and prepared the aircraft for flight.

Inter Mountain Aviation operated the Marana base at that time. Our liaison with them was through a remarkably friendly person known only as Snozz. Yes, he had a big nose but not that big as to warrant his nickname. Snozz greeted us on arrivaland took us to a housing block on the base, which was reserved entirely for us. Each block contained approximately twenty single comfortable bedrooms and a large lounge and recreation room. During later visits, we were offered several four bedroom bungalows one of which was named the de Havlland house. We were then taken to a convenient office location, beside the airfield, in front of which we parked the aircraft. It was a good set up to promote team spirit and get work done.

At six a.m, the weather was clear and calm as we commenced take off and landing tests. Due to the very short take off and landing distances, especially at light weight, the performance camera recorded only a quarter of the image slices available on large glass plates. Vital data, such as the start, unstick, touch down and stop point, could be lost between frames. I introduced a survey theodolite to fill those, often missed events. The unstick and touch down event was identified by focusing the theodolite on a red flag carried by one of three observers, located on the side of the runway, in the expected area of the event. The flagman would hoist his flag marking the location of the event.

By noon, we completed enough tests to keep me busy checking data validity, as I was not prepared to return home and find that recorded data was inadequate for further analysis. After lunch, while the flight crew Mick Saunders and Barry Hubbard, conducted climb tests, I was able to examine photographic records from the performance camera as well as from the photo- panel camera on board the aircraft. As the days went by and the weather was still perfect for testing, I was faced with two problems. Data verification was rapidly becoming a bottleneck and I was also concerned that crew fatigue could cause a serious problem. With clear skies, calm winds and an enthusiastic team, it was difficult to call a halt. I had not foreseen such rapid progress. A quick conference call to the folks at Downsview sent me the help I needed, a data reduction technician. After completing about 75 % of the test program in only five days and well ahead of the four-week schedule, we all agreed a day off was warranted. While the newcomer was busy looking at data, we headed off to visit Tombstone and the desert museum and recognized immediately that we were in cowboy and desert country.

On the eighth day after arrival with all tests completed, we loaded the aircraft and flew to Flagstaff. I could see the frai-looking tower, which had been built for us and soon after landing went to inspect it. I was shocked to see how flimsily it was built. It stood sixty-feet tall and indeed, it was above the trees. Made from standard builders scaffolding, placed in a square and mounted one on top of the other, they were held upright by thin strands of guy wires. There were two platforms at the top. One was for the performance camera and operator, the other I would occupy with a theodolite, a radio and a briefcase holding technical data required for the tests. When I reached the first level I realized, that unless I was a skilled mountain climber or a steeplejack, I could not traverse the overhead extension of the first platform floor. I arranged for the contractor to cut off the overhang on both platforms and to secure a ladder between the lower and upper platforms, which would extend sufficiently over the top platform to allow me a somewhat dignified entrance to the very top. With safety lines in place and lots of onlookers from our team, our photographer, Frank Horton and myself, climbed up and proceeded to haul up the test equipment.

Frank was not a young man and was not far from retirement. When it came time to hoist the heavy performance camera (approximately 100 pounds), it was tough going and our pilot Mick Saunders, scrambled up to give us help. He later vowed he would never climb up that tower again.

Now comes the embarrassing moment. We prepared the aircraft and called for fuel only to be told that the airport only supplied fuel for piston aircraft and no jet fuel was available. At that time there was not only a fuel shortage, but also a shortage of fuel tanker trucks. The president of our company contacted a friend, who happened to be the president of Shell Oil Canada, yet nothing could be done. I remembered seeing a half-ton truck in Marana which had two wing tip fuel tanks strapped to the open truck bed. A quick call to Snozz and the truck was on its way with the small tanks full of jet fuel. The truck had been used to fuel helicopters that had run out of gas and had landed off-road. The driver and truck were at our disposal. When the fuel tanks were near empty, the young driver returned overnight to Marana, topped up with fuel and returned the next morning.

Early morning, after a pre-flight briefing, Frank and I set off to our base of operation, only to find that the tower had iced up due to a cold and frosty morning. We climbed up very carefully and vowed to bring gloves for our next climb. We aligned the camera and theodolite and communicated to the crew by radio that we were ready. The winds remained below five miles per hour and we recorded about ten take offs, after which the winds increased to an unacceptable level for further testing. As the winds increased to around fifteen miles per hour, a left and right oscillation at the top of the tower was noted and was not too pleasant. Due to the high center of gravity, our body movements could sometimes dampen or increase the oscillations. We soon got down and wondered if that dumb tower would last for the rest of the test program.

On one occasion a thunderstorm suddenly appeared and a bolt of lightning struck uncomfortably near by. Being the tallest structure around we were a prime strike target. Down we came like greased lightning.

It seems strange that as Mick Saunders and Barry Hubbard were conducting hazardous testing on board the aircraft with engine cuts taking place during a critical time on take off, I felt in greater danger up in that tower than I ever did in the aircraft during engine cut take off tests.

It became a daily routine that Frank the photographer would lower, by rope, the large photo glass plates in a special container to be developed later. The more tests we conducted the heavier the container. It was near the end of our test program he was lowering a heavy load of glass plates and he called up to me for help. When I got to the deck below, I saw him valiantly struggling to hold on to the glass records of our days tests. He looked pale and complained of chest pains. Fearing a heart attack, I quickly secured the rope holding the glass plates and attended to him. I got him to sit leaning up against the ladder and took his pulse, which seemed normal. My radio was left at the level above, so scrambling up the ladder I contacted our crew for help. I could see the dust flying as two of our vehicles arrived at the base of the tower. I knew it would be difficult getting Frank down. When the rope attached to the glass plates was lowered to the ground and detached, it was hauled up and tied around Frank's waist. I attached the first safety line to him.

If I could not hold on, he could drop twenty feet, and dangle on his safety line. When Frank claimed he was OK, I reluctantly and carefully helped him over the side with one of my hands holding on to the rope that once held those precious glass plates. The rope was coiled twice around a horizontal bar, allowing me to pay it out and secure it easily I hoped if he happened to fall. He got down to the forty-foot level, where he was required to detach his safety line and hook on to the next 20-foot line. Securing the rope with a Boy Scout knot vaguely remembered, I climbed down to help him attach to the safety line that would take him to the twenty-foot level. When he got there Mick and Barry disconnected him and assisted him down. About five feet above the ground, my rope being shortened by being tied around his waist, was uncoiled from the horizontal bar and I was lying face down gripping the rope with both hands. It was then that Frank collapsed in a heap, bringing Mick And Barry down with him. If I had not let go of the rope, I would have joined them rapidly from a great height. No one was hurt and the medical report on Frank revealed no heart problems. It was suggested that his chest pain was associated with a chest muscle strain due to lowering those heavy glass plates.

We only had a few test points to complete the program at Flagstaff. Frank reluctantly agreed it was not wise for him to go back up the tower the next day and instructed me on loading and unloading the glass plates for each test run. He also instructed me on how to aim and follow the aircraft and told me not to forget switching on the high-speed clock timer for each run and then switch it off to preserve the battery.

The tests that day took longer than usual. I spent most of the morning practicing my photographic skills by targeting a car running down the runway. After three runs I lowered the plates for Frank to develop. The results were acceptable and I pronounced that I was ready for the real thing.

There was not enough room for the theodolite tripod to be moved down to the camera platform, so after each of the six remaining landing runs I scrambled up to take position readings. This movement caused the theodolite to lose alignment and it required re-alignment before taking readings of the flag and stop points. The delay required the aircraft to hold the stop position on the active runway for a long time and caused several approaching aircraft to be waved off, much to the annoyance of air traffic control.

We ceremoniously dumped the sandbags, used for heavy weight tests, packed the aircraft and headed home well ahead of the scheduled time allotted. An hour after leaving Flagstaff air traffic contacted us with a message telling us to return to Marana and contact our company. Phone contact revealed that the Australians were about to sign a contract for several aircraft. Due to the fact that the Australian airworthy standards differed from the Transport Canada standards, we were required to undertake further tests. When I was faxed a copy of the Australian requirements, I was surprised to note that requirements for mainland Australia were different to requirements required when operating in New Guinea. The New Guinea requirements were less stringent, allowing reduced take off and landing distances, probably to encourage development and open up small jungle air strips. A volunteer sandbag brigade enabled us to commence heavy weight testing again, and few days later all tests were completed. Before heading home, I took a quick trip across the nearby border with Mexico and bought a western saddle for a song.

The photo on the front cover was taken prior to leaving. I had recently obtained a young filly we named Shandy, named because her colour looked like a shandy drink made from a dark ale mixed with lemonade. Normally, at work, I would plan carefully ahead on every new project, but Shandy was purchased on a whim. I wanted my three daughters to enjoy owning a horse as much as I did many years ago. We had a house built on two acres of land near the village of Wildfield, which is just north of Toronto. Answering a newspaper advertisement, I took all three of my daughters to a farm about five miles away to see the animal, a fatal mistake. We all fell in love with that long legged five-month old filly and purchased it on the spot. After driving home, I announced my purchase to my wife and got a logical response reminding me that I had not prepared to own a horse. Ready or not, I trudged the five miles with with my two older daughters, Linda aged ten and Susan aged nine. We took turns leading her down the road with halter and rope attached. She was high-spirited and frisky but soon settled down on the walk home. When we arrived, my wife held on to the halter and I held my youngest daughter Anne, who was three years old, on the back of the horse and we walked around the property. The beaming smile on her face was very rewarding, even though I was in deep trouble over not being prepared. I soon realized the problem I had let myself in for. I tethered the horse as we went in for a meal. The horse was not used to being tied up and her violent objections, caused us to fear she could get injured. I was due to build a fence the next day and meanwhile, we took turns babysitting the animal. Off course I took the lions share.

The next day it was a race against time building that fence. It was then time for building a barn for Shandy. There were two, twenty foot long barn beams on the property. I hauled these into place dragging them behind my car and set them fifteen feet apart. I then and arranged a barn raising party with colleagues from work. When Stuart, Jock and a couple of others arrived they wanted to see the plans. There were no plans. It was to be a simple A-frame to be erected over those beams, I stated. By nightfall, after some trial and errors, the barn was built and Shandy had her home. When the saddle arrived, the horse was a year old and ready to support the weight of the older girls. When she got used to the bit and saddle, I placed two old pillowcases tied together and filled with dirt and stones and strung them over the saddle. She turned around cautiously aware of the increased weight and a few minutes later she was ready for Linda to saddle up. Soon it was Susan's turn. Breaking in that horse was an anticlimax. There was no bucking bronco display and she seamed genuinely pleased to have the girls on her back. Getting that horse ready to carry precious passengers was a cakewalk compared to the rough ride needed to certify airplanes to carry precious passengers. Soon the girls graduated to riding on an English saddle, entered horse shows and won ribbons in jumping and barrel racing.

A few weeks later we were back in Arizona with further tests for the Aussies. They required aircraft to operate on firm, dry sod strips, so off we went to Stuart in Florida where we found a representative field to measure accelerations during take off and deceleration during braking.

The Twin Otter was continuously developed and kept company employees busy for many years. The aircraft grew in weight and engine power and carrying capacity as it progressed from a series 100 to a series 300 aircraft. Many wheels and undercarriage changes allowed operation on a variety of surfaces. Small wheel for paved surfaces, intermediate and large wheels for operation on semi-prepared and soft surfaces and giant size wheels for operation on the muskeg in northern Canada. We developed skis for operation on snow, floats for operation from water and amphibious floats to allow both land and water operation. Land planes and float planes were modified for use as water bombers, electromagnetic survey aircraft had a large array of cables circling the aircraft from the nose to wing tip to tail. A military version was developed capable of firing rockets. A special version called the 300S was developed for the Air Transit experimental STOL exercise, described later.

All these developments required a varied amount of testing. As "Sods" law would have it, major modifications would often arrive out of season. For example, a float plane version would be ready for tests just a few weeks prior to the winter freeze up. I remember the amusing sight as we taxied out for an initial evaluation on water. My friend Stuart Nicholson was at a mobile wind station beside the iced-up slide ramp which was used for the aircraft to enter the water. He must have stepped on the icy ramp, as slowly and sedately, while standing erect, he slid into Lake Ontario, like a ship being launched and crawled , soaking wet, slipping and sliding back to shore. Another example, was that initial flight evaluations were carried out on grass, using skis.

When these tests were completed, the aircraft was flown to Yellowknife in the Northwest Territories for tests on snow.

Floatplane trials were a race against time to avoid hazardous spray icing as temperatures dipped towards freezing. Porpoising behavior had to be remedied by modifications to the floats. It was an exciting rough ride, pitching in and out of the water during high-speed runs. Tests were often postponed to the afternoon to allow air temperature to increase. During directional stability tests, at large angles of sideslip, rudder lock was encountered. Instead of just releasing the rudder, the rudder pedals had to be forced back to neutral to straighten up the aircraft. This had been anticipated and auxiliary fins had been designed and manufactured as a contingency. These were installed top and bottom on both sides of the tail plane, overnight, and cured the problem.

Multiple rocket firing pods, mounted below the wings, represented another version of the Twin Otter. It was to be an armed coastal patrol aircraft, for use by undeveloped countries to protect their sovereignty. The rocket firing trials were carried out at the weapons range in Petawawa Ontario with Bob Fowler and myself in the cockpit. Weapon testing was new to our operation. When the gun sight was installed in the pilot's position, we had no instructions as to how to align the sight. By guess and good fortune, using laser beams and theodolites we aligned the sight in the same direction the rocket pod barrels were pointing and headed off to the weapons range. Here we were briefed on the Bristol rockets and arming and safety procedures. Arming took place when the aircraft was pointing in a safe direction, in case of an inadvertent firing of a rocket or all twelve rockets. The target was a concrete bunker to be attacked from about a mile away. A helicopter with a photographer, flew alongside. On the first run, a single rocket was fired. It was amazing how rapidly that rocket reached the target. It was a very near miss and even though there was no explosive warhead, the velocity of the impact created sufficient debris to determine the accuracy of the shot. The tell-tale white trailing smoke also indicated the rocket path and was a sight to remember. On the next run we actually hit the bunker much to our and to many onlookers amazement. On the third and final run, we fired two rockets simultaneously, with both arriving close to the target. Orders for an economical military deterrent did not occur, perhaps because the slow moving Twin Otter was an easy target. It was just another exciting test experience to add to my memories.

A STOL demonstration service, using two STOLports linking Rockcliffe, near downtown Ottawa, to a parking lot in downtown Montreal. A minibus service would deliver passengers from downtown Ottawa to the STOL airport at Rockliffe. Six highly modified Twin Otter aircraft were operated by an air service called Air Transit. Their objective was to demonstrate City Centre to City Centre operation. The operation was to include an area navigation and microwave landing system. This was installed to unburden the air traffic control system. In addition to introducing sophisticated navigation equipment, the aircraft was required to meet the higher certification standards reserved for all large transport category aircraft. The seating capacity was reduced to a comfortable eleven from twenty. Modifications to allow improved short field capability were also incorporated.

A computer generated route program was entered into the area navigation system and coupled to the autopilot. The routes were planned to commence turns when immediately above waypoints, distinguishable by landmarks, like a crossroad, communication tower, church steeple or a small clump of trees. Photographic evidence recorded the repeatability and accuracy over each waypoint. When the integration of the area navigation and autopilot was finalized, the accuracy was uncanny, leaving air traffic personnel completely confident in the provided route such that monitoring was not required, thus easing their burden.

A microwave landing system was introduced which would allow instrument landings in poor visibility. The system was low in cost and could be provided at small airports. The advantage of the system was that steeper than standard approach-to-land angles could be selected. The system was also coupled with the autopilot allowing automatic landing with reversion to pilot operation at two hundred feet above ground level. Autopilot system failure investigations, especially when a maximum nose-down signal was introduced at two hundred feet with the aircraft on a steep approach, was quite thrilling and adrenalin producing.

The aircraft was modified, by installing ground spoilers on the top surface of the wing. An improved braking system with anti-skid was installed The spoilers were triggered by the wheels rotating on touch down; they spoiled the wing lift and allowed additional weight on the wheels immediately on landing. This, contributed to better braking. A small section of the ground spoilers could be operated in flight, gently lowering the aircraft if it tended to float just above the runway. A spring loaded throttle stop was introduced, which when the throttle lever was pulled hard back against the stop, would provide additional propeller drag and gently lower the aircraft. The ability to gently lower the aircraft, at will, provide an accurate touch down with reduced scatter.

This Twin Otter series 300S referred to as the super STOL project, was the forerunner of the much larger de Havilland Dash 7 aircraft and was the most sophisticated of the company products to date. It was also pioneered city center to city center operations. A twenty dollar air fare, would transport passengers between the center of Ottawa and Montreal in less time and at less cost than a major airline. The exercise came to a halt after two years of operation, providing a wealth of learning for future products.

CHAPTER 18

With over eight hundred Twin Otters operating in all corners of this earth, it is inevitable that operators encounter some accidents that require investigation. These investigations have provided confidence in the rugged fail-safe design of the aircraft.

In one particular accident investigation, suspicion was cast on a landing flap actuator that might have caused flap asymmetries. I accompanied Bob Fowler to conduct a flight with the left hand flap bolted shut. After take off with both flaps up, the right flap was lowered progressively to its full down position. At each position a series of handling tests were conducted at a variety of speeds. These tests demonstrated that asymmetric flaps were quite manageable and provided added respect for the Twin Otter.

Another investigation took place using the same crew. This time the engine/propeller controls were deliberately miss-rigged, to allow the propellers to reverse, without an increase in power, when reverse was selected. The selection of reverse was a deliberate action and only to be used after touch down. There were suspicions that the pilot involved with the accident had entered reverse slightly behind the idle gate in flight and thus, increase the rate of descent prior to landing. We approached the investigation cautiously, but soon Bob was able to enter into full reverse with gay abandon. The inner wing on both sides, behind the propeller was completely stalled. The outer wing still had plenty of lift, allowing for good lateral control. The aircraft was responding well to all three controls. When in reverse the aircraft sank like the proverbial brick built shit house, but in such excellent control that Bob could move both arms away from the controls place them above his head and turn towards me as to say "see no hands". The docile behavior of the aircraft was so innocuous and recovery so easy that soon we were descending uncomfortably close to the ground and then immediately altering the sink rate to a climb, by moving engine levers forward. Flight testing gives an excellent appreciation of the aircraft's capability. Accident investigation adds to that appreciation.

Demonstration flying is a special skill that is inherent in just a few pilots. Dave Fairbanks was one of those pilots. He was the Director of Flight Operations. He had the amazing knack of knowing what the crowd wanted. I will never forget seeing him demonstrate the Twin Otter. After completing a spectacular demonstration of the short field capability of the aircraft, he turned the aircraft towards the spectators and while stationary, used wheel brakes and skilled use of reverse engine power to bow to the audience by pitching the nose up and nodding, as though to show appreciation for their now tumultuous applause.

It was in 1979, 15 years after the first flight of the Twin Otter, that there was a growing demand to increase the water carrying capacity of the float-equipped Twin Otter Water Bomber. A floatplane was modified to scoop up two hundred gallons of water in each float. This necessitated a large increase in the maximum weight of the aircraft from twelve thousand five hundred pounds to fifteen thousand pounds. Bob Fowler was unavailable due to his other commitments and Mick Saunders had retired.

The Chief Pilot of the Ontario Department of Lands and Forests was seconded to conduct the tests. It was my first contact with Wally Warner who eventually replaced Bob when he retired. Wally became de Havilland's Chief Test Pilot later. Both of these fine pilots received the Mckee Trophy, awarded to aircrew for their outstanding contribution to aviation.

I met Wally in Sault Ste Marie at the floatplane base located in the city. Right away, his enthusiasm impressed me. We discussed the changes made to the aircraft and submitted a compliance test plan to Transport Canada. An initial assessment was made on the narrows between Lake Huron and Lake Superior. The water pick up probes had been modified and water pick up distance was an important feature for an early assessment. In order to operate from small lakes, it was paramount to load up with four hundred gallons of water in a short distance.

A location was chosen on the Michigan side of the narrows to position two transits, a known distance apart, to measure the angular location of the aircraft. Simple geometry would provide the pick up distance. Norman Brittan and Charlie Oliver were the two data reduction technicians who manned the transits. Charlie would later take part in many future remote site operations to provide quick data presentation, immediately after tests. Unfortunately the only location available for those transits was such that the pick up run would bring the aircraft close to some dangerous rapids. We were examining water pick up down wind and down current. I was in the co-pilot's seat and fully expected to pile up on those rocks, on our very first pick up run. With his vast experience, Wally knew the rapid deceleration on water, with power reduction and easily avoided those rocks. The pick up probes were declared adequate, and I headed out to the Canadian Forces Base in Cold Lake Alberta, to negotiate the use of their theodolite tracking facilities at their weapons firing range at Primrose Lake. This would give us accurate information on take off; landing and water pick up distances.

While awaiting for the arrival of the heavy weight water bomber, I was invited aboard a water rescue boat. The boat was standing by for CF 18 fighter aircraft that were to conduct a rocket firing training exercise.

Prior to the arrival of those fighter aircraft, we went fishing. The rescue crew knew the good fishing spots. I have done a bit of fishing in Ireland and in Ontario, but there was nothing quite like this fish story. The fish were large lake trout, the smallest were over thirty inches long. With every cast we caught a fish. The hooks were baited with silver paper obtained from the inner foil of a cigarette package. We feasted on smoked lake trout for lunch and then, came the rocket attack.

Six fighter aircraft flew over in tight formation. One by one they pealed off and attacked a target I could not see from my position, which was approximately one thousand feet from the impact area. It was an impressive display of fire power. The rockets used were Bristol CRV 7 rockets, the same ones we used on the rocket firing Twin Otter.

The test aircraft arrived already loaded with ballast and only needed fuel to get it to a maximum weight and forward center of gravity. The Air Force theodolite crew were available for one hour only, before a bus was due to return all the Primrose Lake personnel back to Cold lake.

We commenced take off and landing measurements. When the hour was up, they offered to stay as long as conditions were suitable. We got a lot accomplished that first day, even though our task was given the lowest priority. With good calm conditions and excellent cooperation, all tests were completed in three days. Worthy of note were, the heavy weight, engine cut, take off test runs. At the new maximum weight, allowed only on the water bomber, a critical engine failure during the take off was required to be demonstrated and measured. Two options were investigated. One was to abandon the take off and come to a stop. The other was to continue the take off and dump the water load when the floats were just skimming the water surface. For both scenarios, handling qualities were quite acceptable and the distance required was measured. On the continued takeoff, the four thousand pounds of water dumped created a sensation of relief as though an unseen hand came to the rescue and slung the aircraft into the air.

I was very impressed with Wally Warner's test-flying capabilities, his enthusiasm, willingness to learn, and his communication skills.

When later, it came time to groom a replacement for Bob Fowler I had no hesitation in adding my voice to recommend Wally as a candidate.

The Twin Otter aircraft ceased production in 1988 when our company's ownership was transferred to the Boeing Aircraft Company. The Boeing Company was uncomfortable with the large number of "unsophisticated operating companies and bush operators that were successfully using de Havilland aircraft all over the world. According to Boeing, the success of the simple, rugged and typical de Havilland products, heralded their doom. This was simply because the new management was not only concerned with the threat of potential litigation in the form of class action suits, they wished to steer all company efforts towards the development of the Dash 8 aircraft. A slow down in the order book, helped their decision to cease production of the older aircraft.

The Twin Otter first flew in 1965 and forty-four years later it is planned to be back in production, to be manufactured by Viking Air Limited in British Columbia. This unusual plan to get back into production says a lot for that much loved de Havilland product.

CHAPTER 19
Survival

There happened to be an unusual period of inactivity in the early seventies, after the Buffalo and Twin Otter development and before the first flight of the Dash 7 aircraft. During this period we were given the go ahead to seek test contracts from companies that could use the expertise we had to offer. The glossy brochure that was sent out to a variety of potential clients even impressed me, which goes to show what a good sales and marketing effort could accomplish. The contracts we hoped to obtain would also serve to maintain the nucleus of our flight test department.

Our first contract was to record hot day performance tests on a Saunders ST-7 aircraft. The aircraft was a conversion of the de Havilland Heron, a vintage aircraft. The fuelage had been stretched and two Pratt and Whitney PT 6 Turboprop engines, similar to those used on the Twin Otter, replaced its four Gipsy Queen piston engines. The aircraft could now carry twenty-three passengers. The aircraft was to be piloted by the founder of the Saunders Aircraft Company, which was based in Manitoba. The Manitoba Provencial Government provided funding for the program. A United Kingdom consulting company was tasked to conduct analysis of the data we would provide. When I heard that the hot weather tests were to be conducted in Spain, I suggested that Marana in Arizona would be a more economical and logical choice. This was agreed to, much to the disappointment of Larry Saunders, who was looking forward to a Spanish location.

Once the tests commenced, it became apparent to me that wheel brakes were inadequate and that the thrust at idle engine power was too great. Financial burdens prevented any changes, so the tests continued.

It became quite routine for the aircraft to end up in the sagebrush beyond the end of a five thousand foot long runway. The overrun was made evident by the dust cloud that engulfed the aircraft as it entered the desert beyond the end of the runway. Clearly my choice of test location was justified.

Communication with the pilot, who was the only crewmember on board the aircraft, was almost non-existent. There was no pre-flight briefing. Complaining, did not provide a solution. Often only the aircraft racing to the take off start point would announce a test. With this as our only indication that a test was to take place, we raced to the camera position and wind and radio station to start recording. Often the winds for the tests were outside limits, or we missed the start of the take off, thus negating the test point. This experience led me to appreciate the test cooperation and excellent communication I had experienced from most of the test pilots I had worked with. It was a frustrating exercise; the recorded data was reduced to an acceptable format and sent to the consultant company. The mission was accomplished somehow, but not to my satisfaction.

Tracker Fire Bomber

Our next contract was with Ontario Department of Lands and Forests. We were required to help them develop a modification of the Grumman Tracker aircraft and convert it to a fire bomber. The Tracker had been built under license by de Havilland many years previously and supplied to the Canadian Navy. I had been appointed as both project and test engineer. As project engineer, I gained valuable experience coordinating the efforts of our design departments, with the modification team of the Department of Natural Resources based in Sault Ste Marie Ontario. The Tracker was originally designed in the United States as a submarine hunter for the Navy. Powered by two Pratt and Whitney radial piston engines. It could get to the site of a new forest fire a lot quicker than a Beaver, Otter or Twin Otter. It coud be loaded with a chemical fire retardant, which could also serve as a fertilizer to encourage new growth after a devastating fire. The Tracker was intended to supplement the existing water bombing fleet. With instrumentation installed to record handling qualities, I flew as co pilot with Bob Fowler to check the feasibility of the aircraft to conduct its new intended role.

With satisfactory completion of this evaluation, we flew the aircraft to Sault Ste Marie for modification. The delivery flight will always be etched in my memory as one of the most enjoyable flights I have ever undertaken. It was early spring and the lakes were still frozen. The surface of the lake was covered with wind streaks of snow and pressure cracks, giving the surface a texture that added to the sensation of speed, as we flew fast and very low over the lake. This was nothing but a pleasure flight and compares favorably with some of the enjoyable occasions when returning from test flights in Northern Ireland. There, we would fly in and out of the glens of Antrim and dance around clouds with nothing to do but enjoy the scenery. These occasions, with no testing to concentrate on, were rare, but necessary to maintain enthusiasm for the joy of flying.

It was early fall when the aircraft was modified for its fire bombing mission and testing commenced at Downsview. The aircraft was now equipped with three fire retardant tanks, each with a rapidly opening dump door. We used water for the tests to check out pitch-up characteristics of the aircraft during water drops. These drops were conducted in increments, building up to a maximum load drop. A final test was to cut the critical engine on take off with a full load, dump the load and continue the take off. Most of the drop tests were conducted over Lake Ontario and we were quite aware of the hazard that could be caused by water freezing. Monitoring water temperature became an essential requirement.

Some spectacular photographs were taken by the company photographer, who was completely drenched in a valiant attempt to get his perfect picture. Next, Transport Canada sent a pilot to verify our tests results. The pilot chosen was Seth Grossmith, who I later flew with on the Augmentor Wing Buffalo research aircraft. All went well and we obtained type approval for the Tracker fire bomber. The Tracker I am sure put out many forest fires and I feel proud of the small part I played in making it happen.

Another contract we obtained was from a large building construction company. We were asked to obtain statistical data to verify the height of aircraft taking off and landing above a proposed large building construction site located below the flight path of a particular runway.

This was an interesting challenge and we came up with a simple cost effective solution. Three four-foot-long wooden poles were fitted with gun sights, allowing them to be aimed at an overhead aircraft. On each was attached a pendulum connected to an electrical gauge which measured the vertical angle of the pole as it followed an overhead aircraft.

The pole "gunners" were spread apart in a equilateral triangle and the positions of each of the pole locations were carefully surveyed. The electric gauges on each pole were connected to a small computer, which by simple geometry obtained the resultant height of the aircraft as it passed by our crude but effective tracking station. We spent three days aiming those lethal looking poles at aircraft flying approximately five hundred feet above. If it had happened after the terrorist attack on 9/11 we would have been arrested as terrorist suspects. That did not happen. A year later, low-rise housing was built on the field from which we conducted our survey.

Next, the Ontario Department of Transport wanted stress loads and vibration measurements taken on a road bridge passing over the, always busy Gardiner Expressway. The information gathered, would give them an understanding as to the life span of that and similar bridges. We dispatched an instrumentation team to install strain gauges and accelerometers and attach them by wires to a recording black box. It so happened that the bridge to which we had dispatched our team, was over the route to be taken the following day by the leader of the Soviet Union. Mr. Brezhnev was due to pass under at noon the following day. Our guys were arrested on suspicion of attempting to blow up a bridge when the Soviet leader was to pass under. After frantic phone calls and the presentation of the contract with the department of transport, the alert Toronto Police freed our men.

A large oil company tasked us with investigating a fatigue problem they were encountering with road signs, which advertised their brand of fuel. The signs were fluttering in the wind, causing metal fatigue at the attachment to a pole. The heavy sheet metal sign would fly off and become a dangerous flying object. It would have taken a long time to find a variety of wind conditions to conduct measurements, and a wind tunnel program, would have been expensive. Instead, we simply placed the pole-mounted sign behind a Buffalo aircraft and used the propeller slipstream to create the necessary winds. Test results , showed that the natural frequency of the sign was excited by a combination of wind speed and direction. The oil company was advised to change the natural frequency of the sign, by installing two signs back to back on each pole. Only fifty percent of their signs could thus be used but they would not become lethal flying objects.

The Commander of the Aircraft and Armament Test Establishment at the Canadian Forces Base in Cold Lake, contacted our Company, seeking support for some military test operations. Len Wise, who was Chief Instrumentation Engineer and myself were chosen to go there and ascertain their needs, relevant to our capability. We arrived in Edmonton in early February. The temperature was minus forty-two degrees Centigrade, winds were high and there was freezing fog in the air. Moist frozen water droplets in the nasal passages are painful and only limited exposure could be tolerated. We were concerned that the car we rented to drive to Cold Lake, on a road with hardly any traffic, would brake down and place us in serious jeopardy. It didn't.

On arriving at the base, we were given a tour of their facilities and test operations and soon ascertained where we could be of help. Help was offered to measure the effect of external changes, such as changes to aircraft performance with the installation of fuel tanks, camera pods and large bubble windows for search and rescue. We also offered to design and install test instrumentation packages.

During the visit, I was made aware of the difficulty they were having with a helicopter project. They were unable to achieve a satisfactory autopilot altitude-hold during flight or in hover. They had tried many electronic fixes but none worked. Though unfamiliar with helicopters, I had come up against a similar problem on the Drone Canberra many years ago and offered to examine the aircraft static air system which senses altitude. My hunch was right, the static air system had been tapped into for other tasks, increasing the volume of the air sensing system. Because of this increased volume, there was a delay in sensing height changes, which created poor altitude-hold. An isolated air sensing source for the autopilot fixed the problem. I use this episode to demonstrate that experience often trumps knowledge.

We were given the contract, but it was withdrawn due to objections that it was a closed, not an open bid contract. The contract was given to a local Edmonton consulting company who later made the lowest bid.

We managed to keep the core of our test team together. It was a satisfying experience in self-preservation and soon we were able to bring back some of our laid-off colleagues and plan the next company product, which was the Dash 7 aircraft.

CHAPTER 20
DH DASH 7

The traditional animal or bird names for de Havilland aircraft went the way of the dodo. Tradition was abandoned and the Dash series began. It was after a meeting of company directors that the news of the go ahead for the next generation de Havilland aircraft was announced. They came out of the meeting all smiles, each with a can of seven up, a straw and a packet of life savers. This was symbolic of the Dash seven, giving the company a new lease on life as the last life-saving straw, in its survival. It was a great day, as we did not need to wonder any more what the future had in store for us. The Dash 7 was designed to operate from short fields similar to those used by its smaller predecessor the Twin Otter but with approximately three times as many passengers. Four Canadian Pratt and Whitney PT6-50 engines driving slow rotating Hamilton Standard propellers, which provided a low noise profile, powered the aircraft. A wing mounted high on top of the fuselage and a high T-tail were distinguishing features.

It was 27 March 1975 when the aircraft conducted its first flight. Bob Fowler was the pilot in command; Mick Saunders was his co-pilot; Jock Aitken and Bob Dingle were the two Test Engineers aboard. I monitored the flight from a radio in my office. It was a two hour and twenty minute flight. What was notable about that flight was that nothing untoward occurred while conducting handling assessments in the restricted speed and maneuver envelope that was carefully chosen for the initial flight. At the debriefing there was standing room only as those hard working designers, were all invited to hear the glowing comments on their design efforts. Their smiles soon faded as time marched on and flying progressed, expanding the speed, weight and center of gravity envelope. It was a start of the thousand modifications, which eventually led to certification. The first ten flights pretty well revealed all the problems we had to deal with. These problems were control flutter, a rudder that was so powerful that it had to be restricted, especially during cruise conditions, unacceptable stalling characteristics, and longitudinal stability discrepancies. Knowledge of these problems gave us the great advantage of responding early in the test program. Important players who responded to many of these handling problems were Dick Batch, Tom Nettleton and Roy Madill, who were key, amongst many others to produce a successful stable of unique aircraft.

When conducting a negative G fuel system test, we got a nasty surprise.

Being prudent, we protected the non-test engines (#1 #2 and #3) with a pressurized stand-by fuel system. When negative G was applied by pushing the nose down in a parabolic maneuver, all the engines (including #4, the test engine) remained functioning. However, all four propellers went into the feather mode, causing a lot of consternation. With all props in feather there was no power. The propeller torque on all the engines went beyond the red line warning on the torque indicators. On recovery to normal G, the propellers remained feathered and manual selection of un-feather caused a very slow return to normal operation. Only low engine power was used to return to base because of the over-torque concerns. We had neglected to realize that on this aircraft, positive propeller oil pressure, was required for normal propeller operation and that loss of prop oil pressure, due to negative G, would cause all propellers to go into feather. The mantra I had always used when planning a test "What happens if ???," did not trigger the caution that could have averted a nasty situation. Needless to say, modifications to the propeller mechanism were incorporated to prevent any further inadvertent propeller feathering.

Soon most development flying was complete and we were ready for remote hot and high airfield tests. Prior to the tests, I went to Marana and Flagstaff in Arizona to set up the test operations. Landing in Tucson on a commercial flight I could not stop the smile on my face as prior to landing the aircraft flew over the familiar site of Mount Lemon. En route to Marana in a rental car, with the, always blue sky and the warm sun above, I could not stop smiling once again, as the distinguishing shape of Pecaso Peak came into view. Evergreen Aviation had taken over from Intermountain Aviation, but Snozz, our previous contact was still there, making negotiations simple. The next day, I drove to the seven thousand foot high Flagstaff airfield to make test arrangements.

The airport manager I had met during the Twin Otter tests, had just returned from hospital after a heart attack and did not want the aggravation of another test program. In a way I was glad not to have to build another high tower for the location of the performance camera.

Nearby was Prescott, which had a five thousand foot high airfield, with runways long enough for our tests. It was dark when I left Flagstaff and stopped overnight in Sedona. When I awoke the next morning and drew the curtains back, I was amazed at the stunning view of bright red rock canyons contrasting with mountain greenery. I walked down the quaint main street to find a breakfast cafe and thought what a wonderful place it was. Sedona today, has grown significantly and is now a great tourist attraction, the pride of the South West.

On arrival in Prescott, I had no problem setting up the test operation, and was able to arrange suitable hangar space. I was unhappy with the height of the airfield, but our Performance Engineering Department was satisfied with my choice. If I had been able to operate from Flagstaff, we could have extrapolated our results without paying any penalty, up to a twelve thousand foot airfield height. At Prescott, extrapolation without penalty, would be ten thousand feet. There were several airfields around the world above ten thousand feet, why pay a penalty? I had an ace up my sleeve. I was looking for an airfield to conduct high altitude engine tests. The airfield in Leadville, Colorado, was nine thousand feet high, but only four thousand feet long. High altitude helicopter tests had recently been conducted there. If our results in Prescott gave us confidence, we could conduct a few critical tests at this high airfield and thus negate any penalty.

With this agreed, I drove on to Leadville and met with a husband and wife team of airport managers who were glad to have us operate there. This completed the determination of the remote sites for the Dash 7 program.

A fifteen man team and arrived in Marana for low and hot teats. Our team met at Little Abner Steak House for dinner where we met up with Snozz who, once again was our support contact. The steak house became a frequent watering hole for many future operations.

When we first encountered that steak house, it was a small isolated outpost in the desert with nothing much around. At night, smoke from a large mesquite-burning barbeque wafted high in the air, heralding the feast that was to come. The place was beautifully lit up at night and looked fabulous. In daylight, it looked like a building that was due to be demolished. Beside the building was a fenced corral with tiered benches. Here, on weekends, a local rodeo was held. Some of the waitresses were bull riders; they were full of life and would whoop and holler at times for no reason at all. Later, with the rapid expansion of Tucson, Abners also expanded and was surrounded by an urban sprawl.

The familiar mesquite smoke still rose lazily into the air as a beacon to hungry diners.

We commenced with airfield performance and climb tests. I was anxious to complete engine and system cooling tests while it was still hot and mid summer. Checking the weather forecasts for nearby locations, Yuma was the most promising with over 105 deg F expected the next few days. The aircraft had special test equipment installed, which was able to load the electric generating system to a maximum design load. This essentially, was a large radiant heater with cooling fans to dissipate the heat inside the passenger cabin. The aircraft had not yet been fitted with any insulation. We got to Yuma midday, with ten aboard. On arrival, we were told not to exit the aircraft, as there was some hush-hush military tests taking place. The control tower was advised that it was essential for our tests that we position a portable air data system, complete with an operator, in front of the aircraft, during engine ground runs. This they reluctantly agreed to.

One of our men was sent outside, in front of the aircraft, a handkerchief on his head and a bottle of water to keep him cool, in the blazing hot sun. He was to relay wind speed, wind direction temperature and humidity, to the aircraft. The rest of us were baking as if in a heated oven. Opening doors and escape hatches did not provide much relief. If a dog was left in a car in the heat of summer, its owner could be charged with cruelty to animals. There I was, responsible for nine people with all of us slowly cooking. I had previous experience with heat exhaustion and heat stroke. Not much longer I thought, waiting patiently for forty temperatures measurements to stop rising, as we ran the engines at a variety of powers. We splashed water over each other and used makeshift cardboard fans to waft each other. Just before I was ready to call it quits, temperatures stabilized, we got our man and his equipment aboard and closed doors and hatches as we rapidly taxied out to the take off point. Temperatures were recorded during the take off and subsequent climb. Hot day engine and systems cooling tests were completed and not to be forgotten.

Back in Marana airfield performance tests were soon completed. Transport Canada, who were responsible for granting Certification, sent their pilot Jake Wormworth to participate and witness the tests. The local workers at Evergreen Aviation were very impressed with our steep angle landing and short ground rolls. These steep landings would ensure plenty of clearance above obstacles around the perimeter of airfields and allow city center operations. After a couple of weeks in Marana, we flew on to Prescott for high altitude tests. Jake accompanied us there. One of the tests that Jake was interested in, was to check the ability of the aircraft to take off, immediately after abandoning a steep landing approach and touch-down. We had simulated that action previously in free air, and Jake rightly so, wanted to conduct the maneuver from an actual landing. I was directing the flight from a radio station atop a small hill, with a good view of most of the runway. This was a simple straightforward task and I had no qualms about it. After descending from a steep approach with forty-five degree(maximum landing flap angle), Jake touched down and slowly and deliberately increased power to go around. In doing so, with such a large flap angle, the aircraft started to wheelbarrow with the nose wheel on the ground and the main wheels high in the air. Directional instability was apparent as the aircraft snaked across the left and right side of the runway, knocking down the frangible runway lights on both sides. The aircraft went out of sight behind a small hill and I fully expected to see a plume of smoke. Thankfully that was not to be as the aircraft reappeared climbing above the hill. The debriefing revealed that Jake was concentrating so much on trying to maintain a straight line, that he was reluctant to increase power further. This would have caused the flap angle to automatically and rapidly reduce to twenty-five degrees and relieve the situation. Mick Saunders, who was in the co-pilots seat, rapidly slammed the throttles to the take off position and direction stability was regained immediately. The problem was resolved by ensuring that pilots would be instructed to rapidly increase power to take off power in a go- around attempt. A second attempt at that maneuver was delayed until damage to the undercarriage was assessed. There was no damage. Those frangible runway lights were indeed frangible.

    The repeat go- around was flawless. The interconnect between the throttle and full landing flap was the brain child of our chief designer, Fred Buller, who recognized that no harm or significant loss of lift would result when flaps were rapidly dumped from 45 degrees to 25 degrees. Only a rapid reduction in drag would help with the go- around. Because of the steep landing capability of our aircraft, Transport Canada introduced a special requirement written for Steep Landing Transport Category Aircraft. We were asked to demonstrate an abuse of a steep landing by landing from a two degree steeper angle than our recommended seven and half degree angle. We had been using our makeshift visual landing aid, known as the "Lolly Pop and Bar", for both three and seven and half degree landing tests. On the side of the runway, a florescent red round disk was mounted on a pole, which was the Lolly Pop. At the correct distance away, was a similar coloured long strip, which, was placed on the ground; this was the Bar. By aligning the disk or head of the Lolly Pop with the Bar, the pilot could be assured he was coming in at the required test angle. For the abuse case, rather than moving the Lolly Pop closer to the Bar, I elected to raise the Lolly Pop higher by drilling an additional hole to pin and raise the Lolly Pop to give us a nine and half degree landing angle. With the abuse case set, I elected to occupy the jump seat between the two pilots, Mick and Jake. As we came in, with maximum landing flap, the descent rate appeared to be a couple of hundred feet a minute more than I expected it to be. We were descending at sixteen hundred feet a minute, and I advised Mick that he should be prepared to rotate and arrest the landing, starting from a higher altitude above the runway, than was normal. There was no need for my advice; the ground was coming up to us so fast that the fear factor alone caused an early rotation and the aircraft touched down gently but firmly. On examining data, it appeared that we landed from a steeper angle than nine and half degrees. On examining the geometry of the Lolly Pop and Bar, It was determined that after drilling the additional hole to raise the Lolly Pop, it was placed closer to the bar than required, thus giving us a landing angle close to eleven degrees. Since we had abused the abused landing, there was no need for any more abuse angle testing. There is always the danger of unforeseen human error in testing, that may or may not cause a serious problem.

While we were busily engaged in the test program, we were advised that a British Oversees Airways pilot and engineer would be arriving and would like to conduct a short evaluation of the aircraft. This was not a good idea as we were still in the development phase of the test program, but our sales and marketing department insisted. When the two-man team arrived we were in the middle of some tests, and we advised them that they were booked into the motel where we were staying and we would meet them there after our days work. Bob Fowler had been sent to fly with them. We had a modest but pleasant dinner at a nearby golf course and conveyed to them what we were prepared to demonstrate the next day. Our flight with them on board apparently went well, and we said our goodbyes as we dropped them off at the civil air terminal. A week later we were sent a scathing memo from the sales department complaining that we did not meet them on their arrival, their accommodation was inadequate and that the dining out was not to their standard. The sales department were politely told not to interfere with a test program again and if the accommodation and food was good enough for us, it should be good enough for the Brits.

CHAPTER 21

It was spring and nesting time for the birds. They may have been starlings and they took a fancy to our aircraft. On most pre- flight inspections, the ground crew would extract unfinished birds nests from entrance gaps around the aircraft controls. There must have been a hidden nest somewhere, as a pair of birds would accompany the aircraft perched on the tail as it slowly taxied out. I guess we lost them on take off. As the aircraft taxied back from the runway, that same pair greeted it, as they hitched a ride atop the tail plane back to the hangar.

There was a day in Prescott I will never forget. It was a Sunday, there was so much activity taking place on and around the airfield, that there was no hope of conducting any tests safely. A glider meet was taking place and glider activity was occurring all around the airfield.

Trailers were being unloaded and gliders were being assembled and towed into the air. Gliders were landing everywhere. At one section of the taxiway, a go-cart racecourse, with safety barriers was being assembled.

Young drivers, boys and girls, fifteen years old or younger, were getting registered for heats, where only the winner could enter the final. The young drivers wore bright race overalls festooned with badges. Spectators were gathering and colourful go carts were being unloaded by proud parents and pushed to the start assembly area. Overhead were about a dozen hot air balloons, low enough that you could hear the blast of burning propane replenishing the hot air. On the ground nearby, a bright yellow balloon was being inflated with a large blower fan. Looking to the left outside the perimeter of the airfield, I could see a group of horses with riders who appeared to be lassoing moving objects, which on closer examination appeared to be tumbleweed. The tumbleweed may have been broken loose by a couple of dune buggies bouncing around some rough terrain. Later that day, I came across a couple of hunters who informed me that they were hunting for peccary, a small pig-like animal. I had seen a family group of these ambling across the runway a few days previously. So much recreational activity, within my eyesight in one day, was a rare sight and made me realize that there was more to life than the daily grind. In my case however, work often represented that days recreation activity. Clearily, my career choice was a good one.

After tests were completed in Prescott, The Dash 7 flew to Leadville in Colorado and commenced engine testing at high altitude. These tests were required to prove that the guaranteed engine power at high and hot conditions could be met with margins in hand. Satisfied that we had the required engine power, and armed with the test results at Prescott, we decided there was no risk to conducting airfield performance tests at an airfield 9000 feet high and only 4000 feet long. The rarefied atmosphere on a hot day at high altitude, presents a challenge for all aircraft.

The Dash 7 had no trouble, conducting engine cut take offs at maximum permitted weight. Light aircraft pilots using the airfield were highly impressed. When publication of the aircraft's performance reached potential operators, our sales team would soon be able to flaunt the aircraft capabilities and make sales in some challenging locations.

To improve The Dash 7's capability further, we went on to other locations to obtain data for rough airfield operation, such as from dirt and gravel airfields. Gravel strips are hardly the same in different locations and tests were required to cover a variety of surfaces with small stones, large stones, round stones, sharp stones, grass and dirt. The desire to operate from airfields other than smooth hard paved surfaces, required improvement in undercarriage strength and the measurement of some critical structural loads. Humps and hollows on rough prepared surfaces had to be tolerated. Surface hardness values had to be established to recommend acceptable soft surfaces. Wheel drag during take off and loss of braking distance after landing was measured. These tests were conducted at various gravel locations such as Chapleau and Earlton in Northern Ontario and also at Moosonee on James Bay, Lake Havasu in Arizona, a grass field adjacent to the runway at Downsview and remote dirt strips in Arizona. These were by no means, exotic locations, except for Lake Havasu, where our accommodations were not far from London Bridge, which had been shipped over, stone-by-stone and then trucked out to it's present desert location. Here, we could ride on an UK red double-decker bus, meet in a typical English Pub and order fish and chips with vinegar, wrapped in a newspaper. With a ploughman's lunch and tripe and onions offered on the menu, it was hard to realize that we were in the hot Arizona desert. We were staying in motel rooms beside the beach and after asking for a suitable meeting room to conduct briefings, examine results and write reports, I was offered a large suite, which I shared with our pilot Tom Appleton. Mr. McCulloch, who was the manufacturer of the chain saw previously occupied the suite. The living and dining room served as the office and meeting room. All window curtains were electrically opened and closed by push- button switches. The view from the windows was stunning. A switch, raised and lowered a large television secretly housed in a beautiful cabinet. A large Jacuzzi bath and a telephone beside the toilet put the final touch to this lavish display of luxury. The night after Tom and I moved in, Tom answered a call coming from my wife, which he passed on to me. My wife appeared puzzled and informed me that the hotel telephone operator told her that I was staying in the honeymoon suite. When the story got around to our colleagues, Tom and I had a hard time with our explanations.

Operating from the gravel runway created a lot of dust, which drifted over a nearby golf course. When the wind came from a certain direction, even though slight, we were forbidden to conduct tests. Those ardent golfers could not be denied hitting a silly ball into a hole.

Tom impressed me with his uncanny knack of touching down in almost the exact spot on each of many landings. I was in the jump seat and noticed a small model aircraft flying beside our left wing tip then breaking away as we reached the runway threshold. We reported this to ground control and continued with the landing tests. After two further encounters with the model aircraft, the encounters ceased. We landed and taxied to our parking area and a police cruiser arrived with the culprit. He was a 14-year-old boy who was practicing his skills with his small model aircraft and tried to keep up with the big boys. After telling him he could have caused a serious hazard, he was clearly apologetic and all smiles as we gave him a tour of our test aircraft. We sent him home with a Dash 7 baseball cap and a large poster of the aircraft. I often wonder if that young lad ended up with a career in aviation.

Early in the gravel test program we had decided to remove the heavy gravel guards that were placed around the nose wheels, after realizing that they served no useful purpose and removal would save weight. Circles made with marker pens identified the areas on the bottom of the fuselage that showed signs of stone impact. The area was identified where chip resistant paint would be applied. To prevent propeller stone damage, all gravel take offs were to be conducted using a rolling take off technique where power was increased gradually until forward movement was detected. Lessons learned on the Air Transit Twin Otter city-center operation were applied to the Dash 7. The aircraft was equipped with a microwave landing and an area navigation system and as with the Twin Otter, the Dash 7 would be able to operate relatively freely, in busy air traffic areas with this equipment. Tests were conducted at the FAA test facilities in Atlantic City, New Jersey. Microwave landings were made from a standard three-degree and a seven and half degree approach angle. The autopilot was to be coupled to both the area navigation system and the microwave landing system. The aircraft could now land in poor visibility down to fifty feet, where visibility would be regained and the pilot would be required to disengage the autopilot and land. One of the special requirements placed on the aircraft was to demonstrate a safe recovery from a nose up and nose down hard-over signal at fifty feet above the ground. This was a standard requirement for all autopilot-equipped aircraft. It was to represent an electric failure that could cause the autopilot to have an un-commanded full nose up or nose down signal. The test had not been demonstrated at fifty feet, especially from a steep approach. To prevent unnecessary risks, I had argued that conducting the test from a safe height and measuring the height loss on recovery, should be sufficient, but that argument fell on deaf ears.

Bob Fowler and Mick Saunders were at the controls and I was operating the simulated failure switch, which was set to apply a nose down hard over signal to the autopilot. The switch was in a mid-fuselage location with a radio altimeter beside me giving me the height above ground. Bob Dingle and Don Band were at the two instrumentation locations behind me. The nose down hard-over from the steep approach was considered the most hazardous and we worked our way down from a "safe" height of one hundred feet, then seventy feet and then down to fifty feet. At fifty feet we were required to conduct many hard-overs to ensure that we covered the worst case scenario i.e. a nose down hard-over occurring when the autopilot had a nose up command just prior to a full down error signal. From my location looking forward through the cockpit windows, we seemed to be hurtling at a great rate into the ground. I gulped as I realized that soon I would be demanding the autopilot dive even more steeply into the ground. Upon recognition of the failure, the pilot had to disengage the autopilot, recover the aircraft and go around, ready for the next approach. We were descending at approximately twenty feet per second. From fifty feet, we would hit the ground in two and a half seconds; this was the time for the pilot to react to the failure, disconnect and recover. The initial tests starting at one hundred and then at seventy feet, gave me a good idea of the response time required to hit the switch at the correct test height. That radio altitude needle, measuring our height above ground, was moving pretty fast towards zero. After operating the switch I just had time to look up and see the difficulty I caused the crew, as the nose dropped down and then thankfully, came up, as Bob recovered the aircraft. In this critical phase of a landing, when the pilot's attention is fully focused on the task, I believe it is the fear factor, causing an adrenalin response, that would guarantee recovery from such an unlikely failure at a critical height. A total of five steep approach, nose down hard-overs, completed the test program. Our baggage had been stored aboard and we were heading home so Mick, who was co-pilot, elected to raise the gear. We had been testing for five hours that day, conducting nose up and nose down hard overs from two different approach angles and flap settings. We had conducted about forty five circuits and elected not to raise the undercarriage after each go-around. The FAA personnel were leaving their offices after completing their days work and since we had completed our last test, we were asked if we could give them a demonstration of our steep and short landing capability. The suggested runway was adjacent to the offices and short sharp turn set up us for the demonstration. A touch down occurred after a steep approach, right in front of a group of spectators. It was a touch down with the undercarriage still up. With a loud scraping sound and the aircraft sitting closer to the ground than normal, we all realized that the undercarriage had not been lowered. Other than that, it was a perfect landing. The wings remained level throughout. As soon as the aircraft came to a stop Bob Dingle disconnected the batteries to prevent any potential fire. Don Band leapt out of the aircraft and held on to a wing tip to hold the wing up and prevent the wing from dropping due to wind gusts. There was no damage to the propellers, as the high wing remained level. In fact, as later revealed, that there was hardly any damage at all, judged by the fact that the aircraft was repaired and ready to fly three days later. The slow aircraft speed used for take off and landing was a bonus; a safety feature unmatched by many other aircraft. This incident was caused, I believe by crew fatigue. During my twenty-eight years at de Havilland, this was the only accident that occurred which caused damage to a test aircraft, giving testimony to the professionalism and skill of the test team and the pedigree of the aircraft.

The aircraft had a powerful rudder, which was capable of producing large angles of sideslip. This feature would allow take offs and landings under strong crosswind conditions. We had only cleared fifteen-knot crosswinds, but the aircraft was capable of much more. Many operators were demanding higher values. To obtain an improved capability, it had to be demonstrated. The critical configuration for cross winds was a light weight and a maximum aft center of gravity. If winds were suitable by the time the aircraft was loaded for the test, the winds had either diminished or changed direction. We kept an eye on weather forecasts and dispatched the aircraft to many promising locations in Canada and the US. All these efforts failed. A Dash 7 operator in New Zealand claimed that his location near Cookstown often had strong crosswinds and offered his aircraft for us to conduct tests. We promptly dispatched a two-man crew, Bob Fowler and Geoff Pyne, who after ten days, never saw a crosswind over fifteen knots. The only excitement they encountered there was a hair-rising ride in a jet boat on a winding river canyon. Not soon after they got back from the failed mission, the tail end of a hurricane reached our home base in Downsview and we eventually achieved a modest increase in cross wind capability from 10 to 18 knots, after traveling fruitlessly, chasing the wind all over the globe. We were capable of a lot more than that, but never got the opportunity of an official demonstration at higher numbers. This subject was now closed.

Another subject to be tested was the evaluation of the cabin and baggage compartment fire warning system as well as smoke evacuation from the cockpit area. A fire on board an aircraft and smoke in the cockpit, are hazards that must be avoided at all costs. A beekeepers smoke generator had been used on the smaller Twin Otter. We decided this time to use a chemical-generated smoke to create a larger volume of smoke. For this, we used a system developed by the National Research Council for use in wind tunnels, to create smoke visualization for airflow over model aircraft. The smoke was non toxic and thus safer for our test crews. Mixing two gasses each contained in a heavy pressure cylinder, produced the chemical smoke. The chemicals, were fed to a welder's torch from flexible hoses where they combined, to create smoke. A preflight evaluation of smoke evacuation and the emergency procedure to be followed, was conducted during an engine run with Bob Fowler, Mick Saunders, Jock Aitken and myself in the aircraft. Jock and I were equipped with an emergency portable oxygen system, which had a five-minute supply of oxygen. Mick and Bob had the quick-donning smoke and oxygen masks, which was stowed in the cockpit. These, were to be used as soon as smoke was detected. All smoke detectors functioned, soon after smoke was released. We were about to engulf the cockpit area with smoke and then Bob was to discharge the smoke by opening an emergency dump valve. They would immediately don the smoke/oxygen masks, and Mick would leave the co-pilot's seat to fight the imaginary fire, armed with an extinguisher. We heard a loud bang and there was no more smoke. The hose supplying ammonia gas, had burst under high pressure, filled the aircraft with ammonia, causing our eyes to burn and ammonia to enter our lungs. We scrambled to don our oxygen masks and get back to the rear entrance door. Jock closed the supply valve on the ammonia cylinder. For a while I was unable to open the rear door, as the aircraft was still pressurized for the test. Bob and Mick in the cockpit had not yet seen smoke, but having smelled the ammonia they had good reason to reach back and grab the smoke masks. After securing their masks, Bob dumped the cabin pressure and both were able to exit the aircraft from overhead escape hatches. When pressure was released, the rear door could be opened and the four of us tumbled out of the aircraft at the same time. The damaged hose was replaced and with a man in charge of the cylinder shut-off valves, we completed the in-flight smoke tests. This incident shows that test hazards are not only restricted to flight tests or aircraft systems. Ground tests and test equipment can also pose problems.

Our showcase for STOL operations was the downtown Toronto airport. It was only used by light aircraft. A Dash 7 scheduled operation from the airport was proposed. No one wants an airport at his or her doorstep. We had spent an enormous amount of time and effort to reduce both the internal and external noise levels of the aircraft. But the general public lump all large aircraft together, as being noisy. When the city fathers gathered at the airport to pass judgment on a potential operation, they were not aware that the Dash 7 aircraft had taken off close to them. Even so, the operation was denied.

The certification tests witnessed by Transport Canada went well. Jim Martin who later led Transport Canada's Flight Test Department, never ceased to amaze me with his ability to home in on areas requiring improvement. Jim Martin, who was a test engineer with Transport Canada, was one of those key persons that were knowledgeable on all aspects of aircraft testing and contributed much to aircraft safety. I had always wished he had been a co-worker.

He and the TC test pilots, during the entire test program provided us with a heads-up on problems they deemed unacceptable. Many large Aircraft Companies have a confrontational approach to acceptance authorities and only reluctantly accepted their role. At de Havilland, we welcomed their early participation, which prevented last minute panics. Jim Martin was given the McMurdy award for his outstanding contribution to Canadian aviation.

One hundred and thirteen Dash 7 Aircraft were manufactured and production of the Dash 7 came to a halt in 1988. The hoped-for city center to city center operation around the world, never materialized in time to benefit the Dash 7. The STOL technology showcased in the Dash 7, was slow to be utilized. Many potential customers using Twin Otter small field operations fell prey to the recession, which coincided with the bad timing of the introduction of the Dash 7.

The Dash 7 saga should not end without reference to our family cottage near Dorset Ontario. I was in a real estate office when a developer came in and mentioned that he had an acre of land for sale on Porcupine lake, which was only a few miles away. It was the only property on the lake the rest of the area surrounding the lake was Crown property and not for sale. The road had not yet reached there, so I had to trudge the last mile on a logging road to see it. Once I saw the land on a small lake, I fell in love with it, rushed back and bought it for a song. A plan had been in the back of my mind, but now it raced forward. The wooden mock up of the Dash 7 fuselage was sitting idle at the back of our hangar and I was determined to use it as a temporary cottage. It would after all, be better than a tent. On enquiring as to the possibility of purchasing the mock up, I was told by our purchasing department that indeed, it was on the disposal list and would go to the disposal company that was the highest bidder. I filled in the appropriate forms and bid five dollars. A week later, being the only bidder, I took possession of the wooden mock up. The mock up was a forty-foot skeletal section of the passenger compartment. It had not been completed due to having an obsolete cross section and had no outer skin. The floor was made from expensive marine plywood, as were the ribs. The first task was to clear out large trees and bush from the property and provide a driveway and the foundation for a cottage, on which the fuselage would be placed. I had plenty of unskilled help, my friend Stuart, my brother-in-law Frank and three of my daughter's friends, Jim, Kevin and Gerald. Jim and Kevin were destined to become my sons-in-law. It took us two days, armed with chainsaws and axes, to complete the clearance except for several large tree stumps, too big and stubborn to remove.

We left with a deep appreciation for the Canadian lumberjacks and the hazards they faced day after day in summer and winter. We were battered and bruised and Frank had a deep gash on his head, from a falling tree branch and had to be taken to the emergency ward in the Huntsville hospital. Removal of the remaining tree stumps would prove to be an easy undertaking when a colleague at work, Milan Kalis, mentioned that his friend could blow up the stumps with dynamite. Milan had recently joined de Havilland. He had worked for an aircraft company in Czechoslovakia and was a refugee that stemmed from the uprising against Russian rule. Stuart, Milan and I arrived at the cottage site and were met by the dynamite man and his eight-year-old son. A hole was dug under each stump and filled with two or three sticks of dynamite, depending on the stump size. A fuse was connected to wires leading to a remote battery. One by one, the stumps were blown to bits by connecting the leads to a battery. The eight-year old boy under his father's supervision, performed this task. We all hid behind rocks and large trees, plugged our ears with fingers and watched bits of bark, tree and rocks fly around. Finally, it came to that very large stump. Four sticks, made no impression, with five sticks, it was the same. When eight sticks were placed under that stump we decided to take safer refuge in a ditch beside the road, which had now reached the property. It was a grand explosion. Bits and pieces were flying high and in every which way. The stump was there no more. The stage was now set to build the base and reassemble the wooden mock up to its new location. The floor, ribs and stringers were assembled in a day. The following day twenty sheets of mahogany paneling were glued and screwed to the frame. The gap left at the top was filled with transparent plastic panels, which gave a romantic view of the stars on a clear night. A front and rear bulkhead with a door and side window encased the Dash 7. Inside dividers produced living, sleeping, washing and cooking areas.

A committee of family and friends designed the outside toilet. We argued about the size of the outhouse, the height of the seat, the placement of the seat and the requirement for a plastic splash guard. The location was determined by our ability to find a spot that would allow a four-foot hole to be produced in that somewhat rocky ground. When it was built it was a joy to behold. Soon after the Dash 7 cottage was built, a township official appeared to assess the "building" for tax purposes. He looked puzzled and enquired if it was meant to represent a large beer barrel. It certainly looked more like a beer container than an airplane. We spent many weekends and holidays in that Dash 7. A heavy snowfall on the fourth winter crushed the Dash 7, ending our unique cottage venture. There were many leeches in Porcupine Lake and my family understandably refused to swim there, so we did not build a real cottage there. Instead we purchased a hundred acre homestead, near the village of Baysville in the Muskoka district. The pioneer home was built in the year 1861. It had once been a sheep farm, stone walls and a tumbled-down barn attested to that. Some time later, it was occupied by a country doctor and a number of medicine bottles were found on the property. The home was in need of much repair, but it served as a weekend retreat and a holiday home.

CHAPTER 22
DH DASH 8

The lessons learned from the Buffalo, Twin Otter and the Dash 7 were applied to the next aircraft, the Dash 8 Series 100. This was a thirty-seven passenger twin turboprop aircraft. With twice the passenger capacity of its predecessor, the Twin Otter, the aircraft was significantly faster and was able to fly a lot further. What it gained in speed and range, it lost in its short field capability. It was designed to be a feeder liner, bringing passengers to major airports from nearby smaller airports. A four thousand foot runway was more than adequate for the Dash 8 to operate from. Such runways were available at many municipal airports.

With great pomp and an elaborate roll-out ceremony, curtains were unveiled to reveal the aircraft to a crowd of workers and dignitaries, including our Prime Minister, Pierre Elliot Trudeau. As the curtains opened in the large hangar, a cloudy mist slowly disappeared with a trumpet fanfare, revealing a brightly painted aircraft.

Two years prior to the first flight on June 1983, a wealth of preparation and much detail planning took place. Many ground test rigs, simulated and tested systems to be incorporated in the aircraft. Fuel, electric, hydraulic, flight controls, and avionics were tested extensively, and many problems were solved prior to installation in the aircraft. Since it was to be an untried engine and propeller on the aircraft, an engine and propeller test rig was assembled in our test hangar and towed out to an area which would reduce as much as possible the noise impact on surrounding homes. Jock Aitken, my colleague from Shorts in Northern Ireland, had a natural bent for engine and power plant development. He was the ideal man to place in charge of designing and operating the engine and propeller test rig. The complex electronic controls of the engine and interface with the propeller were significantly improved and modified prior to the first flight.

A considerable amount of time and effort was spent in providing real-time, on-board visibility of over one thousand instrumented data outputs. The ability to play back test results, without time-delays required for the home base mainframe to spew out data, paid great dividends, during the many remote test operations.

By the time the aircraft was ready to fly, the prior testing of systems and engines had ensured a relatively smooth five aircraft test program.

Bob Fowler and Mick Saunders were in the front seats and Jock Aitken and Bob Dingle were the test engineers aboard for the first flight. As usual, I was monitoring the first flight, but not from my office. There was so much interest from the many hard-working engineers, that communication with the aircraft was established in a large general office. Here they were able to follow the flight and provide me with expert advice in case any serious problems were encountered.

We had an ambitious first-flight test program, which examined many areas of prior concern. At the debriefing, the standard glowing comments that are usually reserved for such occasions, were made and provided a good press release.

Yes, there were problems to overcome during the test program. For example, stalling behavior was unacceptable and the powerful rudder did create directional handling concerns. The problems were rapidly overcome due to lessons learnt on previous aircraft. We had a well-experienced team who were long-term employees. In my career I had not seen such continuity of personnel; it was a great asset not enjoyed by many companies. Experienced engineers once again showed their nettle contributing rapidly to fixes, as they did with previous aircraft, by early recognition of handling problems.

We encountered one nasty situation which was not an aircraft fault but was caused by a lapse in good judgment. The aircraft was conducting Vmca tests, which are required to establish the minimum control speed in flight. These tests are conducted in a take off configuration with the critical engine suddenly shut down. The aircraft then slows down to a minimum speed and demonstrates adequate control. Adequate control is defined as the ability to control a heading within twenty degrees with no more than five degrees of bank. In other words, maintain a reasonable semblance of control. In one of the take-off flap cases, the target Vmca speed was easily achieved with lots of control in hand. The target speed chosen and achieved would not cause any performance penalties and therefore did not need to be reduced to less than that already demonstrated.

With the aircraft showing such exemplary handling qualities, enthusiasm outweighed good judgment and a two-knot lower speed was called for. As the aircraft slowed down towards this lower speed, it encountered a stall with take off power on one engine and with the other engine shut down. The aircraft rapidly rolled over, almost on its back and Bob Fowler struggled to regain control without exceeding structural load limits, which thankfully he did. The crew onboard was shaken and glad to return home in one piece.

This incident was a reminder that test flying is hazardous and that clear thinking is a greater asset than enthusiasm.

Time marches on and Mick Saunders and Bob Fowler were getting close to reaching retirement age. Mick was the first to retire. We had worked closely together on a multitude of test programs and in remote locations. He was a good companion and a perfect gentleman. After retirement, Mick conducted some freelance flight-testing. Several years after retirement, he was in the UK, conducting a post-modification check flight on a Dash 7 aircraft and was accompanied by Bill Loverseed, who I had worked with on the Buffalo and the Augmentor wing programs, had left the company to return to his home country. They were conducting routine engine-out climb tests. They were both well-experienced, test pilots and yet the aircraft encountered a stall when setting up for a second climb in the opposite direction. It is ironic, that between the two, they had carried out thousands of engine out climbs and stalls without incident. The aircraft crashed and both were killed.

Before Bob Fowler impending retirement, a search for his replacement was underway. The successful candidate was Wally Warner, who had demonstrated his abilities on the Twin Otter water bomber program I had witnessed a couple of years earlier. Wally accompanied Bob as copilot, on early development of the Dash 8, he sat opposite him in an office and gained valuable insight into fligh-testing from a master of the art. By the time Bob Retired, he had groomed Wally selflessly to follow in his footsteps and eventually be awarded the McKee trophy, as did Bob for " outstanding achievements in the field of aerospace operations, particularly in pioneering as aircrew, new areas and applications in aerospace". Bob Fowler was also granted the Order of Canada.

When the de Havilland Dash 8, transport aircraft was scheduled to conduct air speed position error tests, corporate responsibility and safety reasons, prevented the previous methods to be used. The highway 400, speed course used on the Buffalo aircraft would result in too many citizen complaints. The low level aneroid method used on the Twin Otter and the Dash 7 was to my mind too hazardous.

During lunch one day, I discussed with my colleagues, the possibility of using a hot air balloon, with a trisponder hanging over the side, to measure the aircraft closing speed as it flew toward the balloon.

They all agreed that the idea had merit, especially as the effect of wind was negated, since both the aircraft and the balloon were in the same air mass. Someone knew someone who owned and operated a hot air balloon. Before long, I had a test plan using a hot air balloon.

It was not easy to coordinate the availability of both the aircraft and the balloon, on a calm day. Two attempts were made to conduct the tests. The balloon pilot, due to wind conditions, that I thought should have been O.K, called off both attempts. My lack of knowledge of operating a hot air balloon, caused me to think that the balloon pilot was being overly cautious. When the third attempt was made, the balloon pilot was reluctant to get up in the air, due to marginal conditions, but up we went. Once airborne, I contacted the aircraft and said we were ready for the test to begin. We were ready, but the aircraft had a minor snag and take off was delayed.

I had loaned my brand new car, obtained the day before, to a young test engineer, Paul Adams, who was to follow the balloon and was to retrieve us after the test. The tests eventually began and it was quite thrilling to see the aircraft fly straight toward us and veer off at the last moment. After the third run, the balloon pilot insisted on landing immediately as the wind was increasing. He picked a ploughed field to which we descended. Now came the biggest thrill. The strong wind dragged the basket, which was now on its side as there was a delay in collapsing the balloon. As we were being dragged, clinging on desperately with equipment flying out, we eventually came to a stop. It was then, that I spotted my brand-new car going through a narrow gap in a hedge, bouncing up and down like a dune buggy, before reaching us in a cloud of dust. I must commend, Paul, that young test engineer who, thought we might have been injured and rushed to our aid. As for my car, with a wash and brush up, it was as good as new. I soon understood the reason for the balloon pilot's reluctance to get airborne. I later found out that he was the person who had been arrested for unlawfully parachuting off the CN tower In Toronto, which was once the highest structure in the world. He was not a faint-hearted person. The results of the three tests points were good, but the balloon test method would take ages to coordinate and accomplish, so this method was abandoned.

It was back to the drawing board to come up with a different test method.

During return trips from many test flights I had often noticed a very tall chimneystack, located in Port Hope near Toronto. I also noted that there was never any smoke from this chimney.

In a light bulb moment, while reflecting on Boscombe Down and that hazardous low racetrack circuit, I envisioned getting an aneroid and camera on top of that chimney. A visit to Port Hope established that the chimney was part of an electric generating station which had never been commissioned and was owned by Ontario Hydro. The caretaker offered to get me to the top of the chimney, which was more than 600 feet tall. This was not a task for the faint-hearted. A small, motorized, two-man cage was to take me up to reach a platform approximately 100 feet below the top of the chimney. It crawled up the inside of the chimney and seemed to take forever. A foot wide rusty iron ladder, 500 feet long, provided an emergency route back to ground level in case the cage motor or battery failed. I could not imagine the terror that would accompany that emergency. On reaching the high platform, a long vertical ladder led up to another platform. Now comes the tricky part. A second ladder takes you up fifteen feet to the lip of the chimney. Looking down these ladders is not advised. In order to conduct the position error tests, two persons would be required to be at the top of that final ladder and a person at the base of the ladder to take notes.

An agreement was made with Ontario Hydro, to use their chimneystack for position error speed tests but only after signing papers absolving them from any liability for accidents that may occur to the test crew, who might be sent up the chimney.

I was surprised that many of our test engineers were afraid of heights, even though they had flown many hours and did not volunteer for the mission up the chimney.

The test method used at Boscome Down was repeated, except at a safe height of 600 ft. Two volunteers went up the chimney with me. One, was to level the camera and photograph the aircraft on each run, the other to log data I would relay to him, with the aneroid and communicating radio by my side atop that ladder. It was a precarious perch. The photographer had no fear of heights and ended up sitting astride on the lip of the chimney. This left me room to place my equipment on the lip.

By flying at 600 feet, safety was achieved for the crew on board the aircraft but provided a hair-raising experience for the ground crew atop the chimney.

CHAPTER 23

When it came time to conduct performance tests, once again Marina, Arizona was the chosen location. There had been radical changes to the air base since we were last there. A prison population occupied much of the accommodation. These were hard-core criminals, mainly murderers, who had served the majority of their time in state prisons. A new experimental policy, placed the prisoners in a prison without walls, a year prior to their release. This was done to aid in their rehabilitation. This experiment was conducted in Marana. Although there were no bars, walls or barbed wire fences, there were many armed prison guards around. We were located in bungalows opposite and across a parade ground from a motel-type accommodation occupied by the inmates. We came in close contact with some detainees during lunch at the cafeteria and when they were put to work cleaning our office area. This atmosphere was too much for one of the members of our team. It was the last straw for him when the trisponder, we were using, mounted on a tripod to measure ground speed, and distance, was found to have a bullet hole which passed right through its casing without causing any damage to the sensitive electronics inside. He packed up and went home. The guards had, it was suspected, used the trisponder for target practice.

Before long, there were four test aircraft assigned to various subjects. The second aircraft was conducting engine and system tests, the third was conducting autopilot and avionics tests and the fourth aircraft had a completed passenger interior and was conducting air conditioning and noise tests. The fifth aircraft entered the test program just prior to certification, to conduct function and reliability tests and passenger and crew evacuation demonstrations.

When searching for a level piece of runway for water ingestion tests, I found great difficulty in finding a location to meet the new standards set by the authorities. A three hundred foot water trough had to be three quarters of an inch deep with a deviation not more than plus or minus a quarter of an inch. After rejecting several local and nearby runways, a promising location was recommended in Chippewa in Michigan, across the Canadian border near Sault Ste Marie. It had an exceptionally long runway, thirteen hundred feet in length. Previously, it had been used by the USAF Strategic Air Command as a heavy bomber base and had recently been abandoned. Thinking we could kill two birds with one stone, I flew out to the base in a Twin Otter with Wally Warner with the idea that it would make an ideal remote but relatively nearby base, for future test operations. The trip was indeed fruitful. Not only did I find the spot for the water trough but I found the new airport management able and willing, at minimal cost, to provide us with adequate office and Hanger space. The long runway provided plenty of test safety and Chippewa became a test site for many ongoing programs. Tests were being conducted at our home base in Downsview as well as in Marana, Litchfield, Flagstaff and Prescott, in Arizona, Santa Fe and Leadville in New Mexico and North Bay in Ontario. The team approach applied to each aircraft, made the coordination of the test programs a breeze. It was only on rare occasions that I was required to visit a team to help solve personal or technical problems. Early participation by Transport Canada paid great dividends and certification was obtained with no last minute surprises.

Improved engines with higher thrust ratings were installed in the second test aircraft and led to the Dash 8 Series 200, which could fly higher and faster than the Series 100. The first aircraft was subjected to a radical surgical operation. Chain saws were used to amputate the fuselage. The nose section ahead of the wing was sawn off, as was the rear section behind the wing. Two inserts were riveted in place, stretching the nose section forward and the rear section aft. The length of the fuselage had stretched, in a few weeks from 73 feet to just over 84 feet. The passenger capacity increased from 37 to 56. The new more powerful engines were installed. The aircraft was given a fresh coat of paint and the tail now displayed large maple leafs sporting the colors of the seasons. It made you feel proud to be Canadian.

Thus the Series 300 prototype rapidly entered the test program and was soon certified.

Many years after achieving certification, a requirement surfaced for the Dash 8 that required operational clearance to conduct Take-offs and landings under high tailwind conditions. Without any testing required, operation in only ten knot tailwinds were allowed and performance data was provided. This tailwind speed was adequate for most operations. No pilot would deliberately elect to operate in a tail wind. A few airfields had high obstacles at one end of the runway such as tall buildings, high cliffs or mountainous terrain. Aircraft were always forced to operate in a direction away from the obstacle. With tailwind components above ten knots, many flights had to be cancelled, which was an annoyance to both passengers and operators. Long term weather forecasts were studied, resulting in Churchill, Manitoba being the best bet for tailwind testing. We were hoping to test in tail winds around twenty five knots. Off we headed to the Great White North. En Route, we flew low over the tundra. It was an unusual sight. The permafrost gave the ground a translucent look as though you could see through the terrain. We climbed higher before reaching Churchill and came across a sunset sky the likes of which I had never seen before or after. Any description that I could conjure up, would not do justice to that scene. All I could say is that it was an out of this world experience.

It was autumn and ice was forming in Hudson Bay, the time polar bears were gathering to commence their winter hunt on the ice. For several days after our arrival, the winds were not strong enough, so we explored the surrounding area in vehicles to look for polar bears. Our first encounter was at the town garbage dump. There we saw four mature bears and several cubs. Moving along near the coastline, many more bears were encountered and we made sure we remained in the safety of our vehicle. Occasionally I would wander out to get a better photo, much to the anguish of the other occupants. We also came across a pure white arctic fox, a beautiful animal indeed. We were lucky to see a ptarmigan, a northern grouse. It blended with the snow beside a bush and was well camouflaged. White winter feathers, just starting to change to summer brown, created a perfect camouflage. It was very hard to spot on the photo I took. knowing it was there, it was still hard to spot. Later, three of us, Wally Warner, our pilot, Jake Wormworth, a pilot from Transport Canada, who was to witness the tests, and myself, took a truck to visit an old missile test range which ran along the coast line. En route, we came across several large polar bears and took great photographs. There was one scene I would never forget. A team of huskies spread about ten feet apart were each chained to a cable. In the Bay about three hundred feet away, was a large polar bear, it was motionless, looking towards the dogs and us. A short time later, the dog's owner arrived with chunks of meat to feed the dogs. We were informed that the waiting bear would not attack the dogs but would wait for days if necessary, to scrounge a few snacks. When the dogs lost interest, as a result of the familiarity of the bears presence, the bear would sneak a meal. The "musher" who owned the dogs, had the look of the North. He wore a padded red plaid jacket and a beaver hat with flaps, loosely hanging over his ears. His boots were lined with what might have been fox fur, which was visible due to the fold at the top of his boots. His face was covered with a bushy reddish-gray beard. There was a rugged look about him, perhaps caused by constant exposure to the elements. His eyes and friendly disposition gave him a teddy bear look. He used his dog team to transport prospectors, hunters or tourists to remote locations in that Arctic wilderness. He left as suddenly he had arrived. We shivered in that cold arctic wind, our parkers not providing the protection as advertised. We then jumped into the truck to warm ourselves and return to town, leaving that patient bear to hopefully grab a snack.

Not forgetting what we were in Churchill to accomplish, when the wind blew we went into action. The winds were strong enough but they were in the wrong direction. This went on for ten days. On the night of Halloween, the whole town was invited to a party given in the brand new community recreation hall. Many of the locals were in fancy dress, and all had a good time. The MC announced they required volunteers to take part in a yodeling contest. One of our gang, who probably had one too many, announced our intention to participate. Hurriedly, seven of us went outside to perform a dummy run. It sounded dreadful. During that venture out in the cold we could hear the sound of random gunfire as the town bear patrol used shotgun blanks to scare off polar bears that wandered into town looking for food. Back inside our pathetic performance received a big ovation. I think it was from sheer pity. Anyway, we were given second prize, a stuffed polar bear. It was late and before we headed to our hotel we were advised to walk in the center of the street so that if we turned onto an adjoining street ,we would avoid any sudden contact with a polar bear.

On the eleventh day that we were in Churchill the long term forecast for high winds was not encouraging, so reluctantly we accepted the fact that it was a failed mission. Armed with souvenirs, such as beaver hats, soap stone carvings, caribou skins and a couple of crates of frozen Arctic char (fish), we headed home. On the way home, we contacted the weather station and were informed that the winds were strong in Northern Michigan. They were also in the right direction for the main long runway in Chippewa, where we had conducted many tests. We arrived in Chippewa and rapidly set up the aircraft and test equipment. The following day we completed our mission with winds remaining high all day. The aircraft could soon be operating in twenty-five knot tail winds, once the performance data was entered in the aircraft flight manual. It was not a failed mission after all.

Times were changing. Bob Fowler retired after a distinguished career as Chief Test Pilot; his legacy was a stable of successful aircraft. He set the tone for communication skills. He was renowned for leaving his door always open to the high and mighty, as well as to the most junior of employees. Wally Warner was now, the Chief Test Pilot, groomed to follow in Bob's footsteps. Wally Gibson, my boss and a manager that any employee would be proud to serve with, was promoted to Director of Technical Design. I lost the title of Chief Flight Test Engineer and was promoted to Experimental Flight Test Manager to take over Wally's position.

Significant changes had taken place to the ownership of the company. In 1963 when I joined the company, the British Hawker Siddeley Group owned it. Dissatisfied with the parent company not supporting Short Take off and Landing development, the Canadian Government purchased the company and shortly after launched the Dash 7, followed later by the Dash 8 aircraft. We had become a Crown corporation in 1974.

In 1986 sales were slumping due to a combination of reversionary times and a disastrous, long Air Traffic Controllers strike, in the USA. This strike caused many of the smaller airlines using Dash 8 aircraft to lose their slots at major airports, causing them to go bankrupt. The company was struggling to remain afloat; the Canadian Government had retained ownership a lot longer than originally intended, and we were put up for sale.

When Boeing Aircraft was announced as our new owners, most employees were overjoyed, thinking that the future of the company was assured. During the first couple of months, there was no sign of changes in management technique. The first development task undertaken by the new owners, was the Dash 8 Series 300A. This aircraft was a modified Series 300, with structural improvements, allowing an increase in all up weight. The main landing gear was moved slightly aft. The weight increase and change to the undercarriage, required a modest amount of testing. The first indication of changes to come occurred during a meeting on the status of the Series 300A test program. At that meeting, it was stated that the time for the undercarriage and flaps to retract and extend was required. This was a simple task, posing no hazard. When I suggested that I contact the aircraft, which was now on its way back from a test flight and get the times recorded, I was told, in no uncertain manner, by a Boeing executive ,that the task was not approved, and in future, all tests were to be controlled by a newly formed test coordination group. All tests had to be approved by the heads of each engineering department. Working through an inexperienced third party with little knowledge of the aircraft, set the stage for micro management, confusion and delays. I was now spending the majority of my time attending status meetings, producing progress (or lack of) charts for the chart room, writing and re-writing schedules and returning test requests after a careful safety audit.

Detailed department procedures were demanded, to acquaint the Boeing management with the way we worked. Many of the Boeing Managers were struck by the range of responsibilities held by the senior staff of the company and noted that the Boeing system would require ten times the number of managers. The three or four meetings a day was attended by such a large number of the staff, that I wondered who was left to do the work. Boeing was trying to change the culture of a small company overnight, to the management style of a large corporation. The simple way of overcoming problems by a visit or phone call to the right person was replaced by committees and procedures. Nothing much was getting done. Experience and knowledge seemed no longer important. I rebelled and refused to attend meetings that I felt were a waste of my time. For the first time in my long career I felt a reluctance to go to work. To give Boeing credit, they did improve office facilities and provided state of the art computer design equipment. They did tidy up the production side of the company, with improved methods and tooling. The Boeing managers were a hard working bunch but seemed unable to change their inbred large corporation thinking. I devoted a large portion of my time to detail planning for the next big stretch. It was to be the Dash 8 Series 400, with even more powerful engines and with the fuselage stretched to almost 108 feet. The aircraft was designed to accommodate one hundred and seven passengers. It conducted its first flight on Jan 31 1998, long after I retired. Wally Warner and Barry Hubbard were at the controls. The test engineers aboard were Angelo Susi and Dave Monteith.

In September 1991, together with many of my colleagues, I took early retirement at age 63, to start a new chapter in my life.

Prior to the day set for my retirement, I received a couple of interesting phone calls. One was from a Boeing pilot who had spent some time with us as Director of Flight Operations. He was calling from China and wanted to know if I would be interested in training and advising employees of a newly formed Aircraft Company there, to test and certify their twin-propeller turbine engine aircraft. He pointed out that Boeing had no experience with propeller-driven aircraft. Without much hesitation, hoping for a new adventure I said I would and started writing lecture notes. A week later, he called back and stated that due to political reasons, the China project had to be abandoned.

The other phone call was from a headhunter who wanted me to go to Sweden with him to aid him in negotiating a contract to supply engineers to the SAAB Aircraft Company. They needed help to certify their new aircraft, the SAAB 2000 which was also a twin-propeller turbine aircraft, due to fly in six months. I joined the headhunter on a trip to Sweden. He was armed with the resumes of many of my colleagues, reflecting the unrest created by the Boeing ownership. I thought after a meeting with SAAB, that such talent would be snapped up. Lockheed Aircraft personnel were granted the contract and now, unexpectedly, I was well and truly retired, or so I thought.

CHAPTER 24
Retirement. (1991 to 20??)

Apart from the Boeing take-over of de Havilland, there were many significant changes taking place in my personal life. In 1985, our 32 years of marriage ended in separation and divorce. My dedication to work was never understood, and periods away from home did damage our relationship. My wife Brenda, during the last eight years we spent together, suffered from mental illness and was in and out of hospitals several times a year. There were three suicide attempts which were most likely a cry for help. Secret alcohol consumption did not help matters. I was unable to cope any longer. I had learnt to cope with and fix many aircraft problems with lots of professional help. This problem, I was not able to fix, nor did the medical professionals provide much help. The first twenty-four years of marriage, we would both agree were pretty good. She was, and still is, a wonderful mother to the girls. The last eight years of our marriage, I would rather not recollect. Separation and divorce was the best for both of us.

I remarried, a couple of days before my retirement, to Myrna, a lady I met when taking an art course. I thought that painting would occupy me after my retirement, it did not, but many other ventures did. I tried line dancing, square dancing, and Tai Chi, none of which I could master as my brain connection to my feet was inadequate for the task. I kept myself busy with dabbling with a few gadgets.

Myrna was a keen alpine skier. The ski club she belonged to, sponsored a group ski holiday in Zermatt, Switzerland. I elected to go along. Before doing so I thought it would be a good idea to try my hand at skiing. It did not go well, although I did enjoy hurtling down, out of control, until I fractured my tailbone on a chair lift. Driving home was a nightmare, as rear muscles I was never aware I owned, went into violent spasms. A couple of days later I was off to Switzerland. The journey, by plane, bus, and train, was no joke. The nest made from my wife's fur coat, that I sat on did not ease my pain, so I spent most of the trip on my knees, facing aft. After resting for two days, I donned my skies and accompanied my wife on some easy?" runs. After somersaulting out of control, I agreed with my wife that lessons for me were in order. My instructor was an eighty-year-old Swiss mountain guide who had escorted many mountaineers to the top of the Matterhorn. The distinctive peak of the Matterhorn was always visible no matter where we were. Three days of lessons improved my confidence, but did very little to improve my capability. Leaving the nursery slopes, I headed for some of the easier intermediate runs that my wife had scouted out for me. It was not too difficult until we came in sight of the restaurant, where we were to have lunch. As we paused, before a long and steep downhill slope leading towards the restaurant, I was advised to build up speed to allow me to climb the steep upward slope that led to the front door. I headed down, building up speed fast, not attempting to weave to the left and right. Later I was told that a rooster tail of snow trailed behind me. Almost falling at the bottom of the down slope, with one ski up in the air, and completely out of control, I streaked up the steep up slope and only came to a halt after I passed through the doors of the building. It was a Kramer-type entrance as depicted in the Jerry Seinfeld TV comedy show. I was a danger not only to myself but to others also and vowed never again to ski out of control. I returned to Canada with a newfound passion for skiing.

The property I had owned near Dorset was sold and my home near Toronto went to my first wife. The old, one hundred acre homestead near Baysville was to be my retirement home. It needed a lot of work to make it habitable on a permanent basis. The large country kitchen/dining room was a somewhat recent addition to the old two story board and batten house. This addition was in the worst state of disrepair and had to be torn down and rebuilt as it was. The tree trunk beams and joists, placed on a stone foundation, were full of wet rot. One could pull out chunks of wood with their bare hands, causing one to wonder why that the house was still standing. A local builder worked wonders, and after major changes, still maintained the character of the old homestead.

The derelict two-hole outhouse was ceremoniously set on fire and replaced with the committee-designed one hole outhouse salvaged from the property near Dorset where the Dash 7 fuselage once was. We spent the first three winters without any indoor plumbing and fetched drinking water from a spring about twelve kilometres away. Water for other use came from rain barrels or by melting snow over the wood stove. Bathing took place using a galvanized steel tub in front of the stove. Towards the end of the third winter, I began to realize the significance of the second hole of the twin hole outhouse that I had burned to the ground. During a cold winter, when the temperature could drop to as low as minus forty degrees, which is the same value Centigrade or Fahrenheit, turd freezes and before long a cone shaped column rises up from below, stretching up towards the toilet seat. Transferring to a second hole, which now did not exist, would have been an option. Those old timers knew what they were doing as I began to realize the significance of that second and sometimes third hole in pioneer outhouses. Boiling Water was the only smelly solution to the problem. An axe could not be swung in the confined space available. Another lesson we learned was that a wooden seat was significantly better than a plastic one, due to the difference in the temperature transfer rates. When inside plumbing and a approved septic tank was installed, we missed the starlit night sky viewed from the outhouse. We lived the pioneer life the first three years and surprisingly, we both enjoyed it.

When the Canadian Ski Patrol was looking for volunteers for the nearby Hidden Valley ski club, my wife and I both applied. Thinking we might be considered too old, we asked if there was an age limit and we were promptly told that if we were over eighteen years of age we were in. So we joined up and I was surprised that I passed the test required to evaluate my ski ability. The first aid course was extensive and provided valuable knowledge for use in any unforeseen emergency. The free ski lessons the ski patrollers were given were welcome and soon I considered myself a good skier.

Every year, the ski patrol sponsored a dummy race. The race entertained large crowds, who greeted spectacular crashes with loud applause. The dummies, mounted on skis, reflected the imagination of the many sponsors of the race. A local bank entry was a shapely and attractive mannequin wearing only a bikini. Almost every year it was a winner. The distance traveled before crashing determined a win. With no directional control, all the dummies eventually crashed. When asked to produce a new dummy for the following years race, I jumped at the chance at using my aeronautical experience. Having noted that directional control was lacking and that dummies with a high center of gravity, were the first to fall I came up with the idea of a direction controlled dummy with a low center of gravity. A model aircraft control system was purchased to provide remote control signals. Two auto door lock solenoids purchased from an auto wreckers yard were mounted on each ski. The solenoids triggered the brakes on the skis, which automatically deploy when a wayward ski comes off the boot, quickly stopping the ski and preventing it from becoming a uncontrolled weapon. A dry cell 12-volt battery was attach to a link connecting the two skis. This supplied power to the solenoids. Two remote control servos were each mounted near the brakes. When signaled, the servos rotated a cam, which then triggered a micro switch, which was wired to supply battery power to the brake solenoid, thus operating the brakes. With both remote control switches forward, brakes were released, propelling the dummy forward. With both switches aft, brakes were powered down to slow the dummy. With the left switch applied only, the dummy would turn left. The right switch only would cause a right turn. Thus, we would have directional control.

The center of gravity problem was solved by the weight of the control equipment being low on the skis and the fact that the dummy was a hollow plastic snowman that weighed very little. A ski patrol scarf and toque adorned the snowman and a small flag with the distinctive emblem of the Canadian ski patrol, masked the antenna. The dummy was completed early summer and static runs proved out the mechanism. It was a long wait for enough snow to conduct operational trials. With the first snowfall, and long before the ski hills were open, the dummy was taken on a test run. All worked as planned and the dummy race near the end of the ski season was awaited with great anticipation. When the day finally arrived, I secluded myself on high ground above the crowd of onlookers and steered the dummy as far as the law of gravity would take it, making the ski patrol entry a hands-down winner. We had an unfair advantage since the rules for the race did not specify that remote control of the dummy was disallowed. Winning the following year I felt that winning was too easy and withdrew from future races. On request, I did demonstrate the remote control dummy at a couple of ski clubs, by skiing in formation with my pal the dummy. After three years patrolling at the hidden valley ski club, my wife and I were asked to volunteer to cover a nearby small ski hill attached to a beautiful resort lodge. Many of the patrollers were reluctant to operate there because the resort catered mainly to gays and lesbians. Due to insurance requirements, the permit to allow skiing at that resort would be withdrawn unless the Canadian Ski Patrol was present there. We volunteered for the weekends and when called upon during the week. It was an easy assignment and the free lunch we were given at a gourmet restaurant was most welcome. There were not many skiers on that small but steep hill, and our first aid skills were not often required. When no one was on the hill we relaxed in the large wood structure lounge with picture windows revealing beautiful Christmas card scenes. Occasionally, we got into deep philosophical discussions with some interesting characters. Knee injuries were a common occurrence and before long a suspected fractured knee had to be attended to. While my wife completed the bandaging, after we had padded the knee in a bent position, I fetched the rescue toboggan and soon he was in the lodge awaiting the ambulance to arrive. He was a bit of a wimp and complained continuously about the pain he suffered. A few weeks after the incident, the resort was sued for his pain and suffering. My wife, being the attending patroller, was asked to file a report. We heard no more about that incident.

Although we had a lot of land, just an acre around the house had been cleared and required the grass to be cut. At first a push gas mower was used. To ease the chore a ride-on mower was purchased. It could turn on a dime. Steering was by two leavers, both forward to go ahead, both aft to go aft. With the left lever back a left turn was initiated, if the right leaver was moved forward at the same time a sharp turn within a wheels length could be conducted. The opposite action was required to turn right. It sounds complex but it was instinctive and simple to use. It did not take long to realize that this mower was a perfect candidate to transfer the remote control mechanism used on the ski dummy and adapt it to the mower. The door lock solenoids were replaced by electric motor driven, window wind-up levers. I now had visions of a hot summer day, reclining in a deck chair with a cold beer and mowing the lawn at the same time. With this tantalizing vision I set to work. A motor and driving linkage was installed on the floor of the mower in front of each of the two control levers. Electric power was obtained from the existing mower battery. The remote control mechanism was mounted on a wooden circuit board in front of the mower, which also supported the antenna. It worked, but there were a few pitfalls. The motors were very powerful and when any command from the remote control switches was made, there was instant response. A turn was so sharp that it was difficult to stop the turn in exactly the direction you wished to go. This often resulted in turning rapid pirouettes and trying to decide when to release control to head in the right direction. Reducing power to the motor improved the problem somewhat. There was also a sense change that occurs when the mower heads towards the controller as opposed to away from him. I attempted to relax in a deck chair with a cold beer; it was not possible. I needed two hands and an alert stance, and could not achieve my vision. There were times when due to a combination of high vibration and crude wiring instillation, a wire would come loose and the mower would run amuck. On one occasion, it headed towards a steep incline only to be rescued by my frantic dash to take over the controls. On a couple of occasions, the wayward mower banged into solid objects, a large rock and a picnic table. The front mounted control circuit board took a beating and required repair. It was a poor design to install sensitive equipment in a vulnerable location. I was considering controlling a selected area using infrared sensors to trigger a short time delay, thinking that it might control the turn. Further development ceased when an unusually intense storm demolished the shed that housed my project. Damage to the control was irreparable and thus my challenging project ended.

When I retired up North, in the Muskoka district, at first our mail had to be picked up from the village post office a couple of kilometres away. Two years later, mail delivery was offered and a mailbox was required to be installed at the end of my two hundred and fifty foot long driveway. Mail delivery times were erratic and since the flag on the mailbox signifying the arrival of mail, was not visible from the house, I got myself a new project. I would conjure up a mail arrival alert signal. There was to be no more trudging in rain, sleet, or snow to find out that the mail had not arrived. My first solution was simple and worked reliably. A magnetic proximity switch, used for a window burglar alarm, was installed to indicate when the mailbox door was opened to receive mail. This signal was sent by wire to a miniature electric motor powered by a flashlight battery. The motor, when triggered operated a lightweight red flag. All this was installed in a small quarter scale model of a mailbox and located near the telephone in the house. Off course I had to improve this. Soon a light indicated that the system was armed and battery power was available. Arming took place when the main mailbox flag, in the down position, set a micro switch. This would eliminate nuisance alarms when retrieving mail. Next a flashing light and a dingdong bell was installed, thus, mail alert was available no matter where we were in the house. As a further refinement a miniature voice recorder announced, in my wife's voice, " The mail has now arrived, kindly retrieve your mail". A friend and colleague of mine, King Chu, had recently retired. He had headed the instrumentation department and was a wiz when it came to electronics; not only did he help me with this project but he produced a professionally-looking system as compared to my usual-lash up arrangement.

When something works well, it should be left alone, but no, I had to make a further refinement. The long wire connection was not practical as mailboxes were often located, across the road, making a hard wire connection impossible. Since I was considering patenting the device, I thought a wireless system would be most appropriate. Experimenting with wireless doorbells, I found that none on the market had sufficiently long range. I placed them in series to solve the problem.. The main mailbox door triggered the remote button switch, which in turn triggered the receiver bell, which was halfway down the driveway. The bell was removed from this receiver and the signal was relayed by wire to an adjacent second remote button switch, which in turn triggered the bell located in the house. A signal from this bell, would eventually trigger the mail alert system already in use. Hope you got all this. Eventually I got it to work but often, bells would be ringing, lights would flash, the flag in the house would go up and down, and my wife's voice would start and stop. A frantic rush to locate the problem and stop that continuous, annoying alert, resulted in the blame cast on the relay half way down the driveway, which formed a closed loop. A short while later, a wireless mail alert came on the market for a modest price. I purchased one; it did not have the range I required, or the bells and whistles I had created. I reverted to my original hard-wired system and gave up on the attempt to gain a patent.

Four years after retirement, I would be back in the saddle, testing aircraft once again. In the early summer of 1995, Joel Paul of Bombardier Aircraft contacted me, requesting that I take part, as a flight test consultant on a Dash 8 test program in Arizona. Bombardier had replaced Boeing as the owners of de Havilland. Joel had been given my position as manager, after I retired. He informed me that remote trials were running well over the scheduled time and were not going well. Any feedback as to why progress was slow would be appreciated. A new requirement to obtain narrow runway certification was to be met and he asked me to organize these tests, as apparently, I had gained a reputation for getting things done. After spending a week at my old Flight Test Department at Downsview, I studied the task and put together a test plan for narrow runway certification.

The task in hand was to certify operation from a 14 metre (46 foot) wide paved as well as an unpaved gravel runway. The main wheels of the Dash 8 were 26 feet apart, leaving only 10 feet to spare on either side of that narrow runway, when the aircraft was exactly centered on the runway. The existing take off speed of the aircraft was such that, if an engine failed during take off, the aircraft could deviate as much as 30 feet. This was acceptable for normal runways. A narrow runway was a different matter and a significant increase in take off speed would be required to remain in the confines of a narrow runway with a sufficient safety margin in hand. The safety margin was defined by a formula based on the width of the runway and the width of the aircraft main wheels. This resulted in the requirement to demonstrate no more than a 6-foot deviation after an engine failure. The demonstration and recording was to be conducted on a runway painted to simulate a narrow runway and was to include continued as well as aborted take offs after an engine failure. Landing with an engine out was also to be recorded and demonstrated. The required increase in take off speed would cause an increase in take off distance.

With the narrow runway test plan completed, I headed out to the test location at the Williams Air Force Base, near Phoenix. The Air Force had recently vacated that large base which was opened for civil operation. Here, tests to certify an increase in engine power and the resulting improvement in climb and take off performance were already under way. There were many new faces there and a lot more personnel than I expected. There had been a ban on overtime throughout the company and, as a result, the large ground crew did not have the aircraft ready for flight until past midday when turbulence and high winds often prevailed, preventing testing. The calm winds, often occurring in the late evenings and early morning, which was an ideal time for tests, but could not be taken advantage of, as test operations started late and ceased after 4.30 pm. Any small changes to test detail plans, which were often necessary, caused significant delays as authorization signatures from many senior engineers back in Canada were required. A lot of personnel had wives and girl friends with them. It was no longer the short-term small remote operation that I had introduced many years ago, but a long term, large-scale operation. These were my findings, which I conveyed to senior management. It succeeded in getting the overtime ban lifted, to take advantage of suitable weather during remote operations.

The airfield authorities at the Williams base were reluctant to allow a runway to be painted to simulate a narrow runway, even though I recommended the use of temporary tape used by the roads department. Off I went to Marana, and when I saw the familiar terrain, I lit up in a smile, brought on by happy memories. At the airfield there, were many stored large aircraft, aligning the runway. The aircraft were parked too close to the runway to be used for the task we had in hand.

Just ten miles away, was the Tucson Aero Service Center, operating at the Avra Valley airfield. They had just expanded the runway to an acceptable length of six thousand feet and were willing to allow the painting of a simulated narrow runway.

The aircraft to be used for the tests was a Dash 8 Series 100, which had been leased from a West Virginia airline. Two flight service mechanics were to accompany the aircraft. The folks at Avra Valley were most co-operative. Knowing that the test aircraft was due to arrive in two days, they issued a "Notice To Airmen" describing the new additional runway markings. As sufficient office accommodation was not available for our use, an office trailer, complete with phone service, was set up and the width of the narrow runway was identified by yellow paint.

All was ready in time for the arrival of the test aircraft. When the aircraft arrived, a computer was attached to the digital crash recorder to provided engine power and airspeed data. A water-trailing hose was installed to provided a liquid marker, to measure aircraft deviation across the runway. There were many familiar faces in the small test crew, that arrived from the Williams air base, where tests had been completed. Bob McKenzie as co-pilot, accompanied Wally Warner, the Chief test pilot. Bob was one of several de Havilland test engineers who moved into the front seats of aircraft and became test pilots. Barry Hubbard, Don Band and Dave Martin were the others. Dave was unfortunately on board a Challenger aircraft and was killed during a fatal test flight accident, after he joined Canadair in Montreal.

Landings with an engine out were a piece of cake, with maximum deviations less than two feet. Aborted take offs or accelerate stops, after an engine failure, also produced innocuous deviations. The continued take offs after an engine failure, required many trials and errors but, eventually the speed chosen provided repeatable demonstrations of deviations, never more than 4.5 feet. The water trail spotters were kept fit by running to the centerline of the runway to obtain measurements before the trail vanished due to the hot and dry desert weather.

The tests went well; I however was not doing too well and limped around with a painful and severely swollen right leg. After a visit to a doctor, a couple of aspirins were perscribed and I was to return in three days. During those three days, the tests on pavement were completed and I went horseback riding at a nearby ranch. The following day, on my way to the doctor's office, I picked up a line marker used for marking a football field grid. This was to be used for marking the narrow gravel runway. As soon as the doctor saw my still swollen leg, he placed me in a wheel chair and wheeled me into the hospital across the road, suspecting a deep vane thrombosis i.e. a blood clot. Left in the supposedly good hands at that hospital, my Canadian medical insurance was not accepted and I was told to find my way to Saint Joseph, a Catholic hospital, where my insurance would be accepted. Exiting the hospital from a different door, I lost my bearings and wandered around trying to find my vehicle, which was at the doctors office. During that search I was aware that I had been told to keep my foot elevated and to remain seated. Finally at St Joseph's, a four inch blood clot was identified. I had plenty of team member visitors. My wife Myrna, flew over from Canada. Jock Aitken, my colleague who was with me in Ireland, took over from me and the gravel test operation was completed. The test aircraft returned to West Virginia. After eight days in Hospital, I joined the test crew and we all returned home by commercial airlines. They returned to work and I returned to my retirement, but not for long.

CHAPTER 25

It was a year after my return from Arizona that I was given a contract by Bombardier to act as a test consultant, once again to conduct narrow runway certification tests, this time on the larger Dash 8 series 300 aircraft. A phone call to Avra Valley indicated that the runway had been re-paved and the narrow runway markings we had painstakingly applied with spray paint were no longer there. By the time we arrived in Arizona with the test aircraft, the Arizona highways department had marked the narrow runway with temporary yellow tape.

Wally Warner and Bob Mckenzie were again in the front seats; most of the other members of the test team were new employees. Tests were well underway on the narrow runway tests when I received a phone call from Joel Paul, the Flight Test Manager. He informed me that priorities had changed and asked me to search for a desert strip, which had a load-bearing classification softer than any previously tested. Apparently Saudi Arabia wished to purchase several aircraft to support oil explorations. They were to operate from low load-bearing strips. While awaiting the penetrometer, which was required to measure the hardness or softness of the soil, and doing some homework on possible locations, we completed the narrow runway certification tests. I was now able to devote my time to finding a soft gravel or dirt strip. I had learnt that there were many private desert strips in the State. Free of charge, the Tucson Aero Service Center provided us with a light aircraft and pilot. Wally Warner and I scoured the desert searching for a suitable long runway. Eventually we found an area about twenty miles from Sun City, which appeared to fit the bill. The pilot would not land there, saying he had not been given permission. I took a mental note of the location after following a dirt track that passed through a dry gulch and an abandoned corral, eventually ending up in a field beside the highway between Tucson and Phoenix.

Early the next morning, I traded my rental car for a four-wheel drive vehicle and headed out alone to look for that dirt road. The others were busy checking data and writing reports. The aircraft test instrumentation was reconfigured to ensure that data required for the new test was available. It was not easy locating that field off the highway and the trail leading off it was hardly a trail at all but I took it anyway. It had rained overnight, the many puddles I ploughed through attested to a heavy downpour. Following that often ill-defined trail through the desert, I eventually found the corral, which confirmed that I was on the right trail. That dry gulch, I had noted from the air, was now a small river. Convinced that my four-wheel drive would get me across, I attempted to press on, I took the steep slope into the water at an angle and was lucky not to roll over into the water, which averaged two feet in depth. Once in the water, I became well and truly stuck. The more I attempted to extract my self, the more I got stuck and eventually I became completely immobile. I sat, scratching my head and contemplated abandoning the vehicle and continue on by foot to find help, when two pick up trucks arrived on the opposite bank. They were hunters out to shoot wild boar and were kind enough to get me out of my predicament. The first attempt was for all four of them to push but that did not work. We were all covered from head to toe in mud. I thanked them for their futile efforts and expected them to carry on with their hunt. Those hunters were dogged and determined to see me on my way. Ropes attached to both pickups hauled me out like, a downed steer, out of that wet gulch. It took almost two hours,for those good Samaritans to help me out and I was indeed thankful. I went back to that muddy river, washed my muddy face and headed on to find that airfield, which I did.

There were some signs of human activity at the end of one of the two airstrips. A drill rig, trailer, and a pick up truck, could be seen. Arriving at that location, I came across a character, the likes of whom I had never seen. The word renegade describes him. He looked dirty, but then so did I, still covered in wet mud. His unkempt bushy hair and toothless grin were accompanied by giggles and slurred speech, which I failed to understand. Beside him, precariously resting on a rock, was a large jug of liquor, which he occasionally took a swig from. He was thoroughly drunk, and I was beginning to think that my ordeal in getting there was for naught, when the small trailer door opened and a woman with long black hair and the look of a Native American, came out. She informed me that her man was a drill digger and had worked very hard for many days and had taken the day off. The good news was that the owner of property was due to arrive by air shortly. While I waited for the owner, I was offered coffee and cookies and watched them light cigarettes by short-circuiting a car battery. Their supplies were dwindling and they had long since run out of matches. The woman was quite good looking and obviously well educated, she seemed quite a contrast from her man, but then, I had not yet seen him sober. A small aircraft arrived flown in by the owner. He was middle aged with a weather-beaten face sculptured by the Arizona sun. He was enthusiastic about our proposed tests and refused to charge for the use of his airfield. He explained that it was a dream of his to start a colony of people who would commute to Phoenix or Tucson by air. He was planning to build homes with a small hanger adjacent to each home, which would house both a car and a light aircraft. He was in the process of digging wells for water and building a show home and hangar. During further conversation, I gathered that there was a feud with a local Indian tribe ,who were not enamored with his vision of the future of the property. After discussing our requirement for measuring the load-bearing classification of the runways, he welcomed the task, the outcome of which would also be useful to him. My next task was to improve the passage through that dry/wet gulch and to obtain a heavy truck to be used to provide the reaction force for the penetrometer used for runway load-bearing measurements.

Before heading back, and not wishing to encounter that wet gulch again, I enquired about an alternative route. The location of a movable barbed-wire fence was pointed out and that the trail from that fence would lead to an Indian village,where directions could be obtained to reach a paved road that would eventually lead to the main highway. A casual statement indicated that the trail out was rough. Off I went, searching for the removable fence and found one at the end of the trail I had used to get to the airfield. Opening the fence and struggling to close it, I headed out into the desert. Soon there was no trail at all, I tried to turn around but a 180-degree turn was not possible. Attempting to turn back I realized that I was lost. Eventually I found my tracks and retraced them back to the trailer and discovered that I had opened the fence where no gate was intended. I was eventually led to the correct fence location. Daylight was soon to be over and that rough trail was a lot worse than I had imagined. It was passed midnight when I returned to my hotel. One look at myself in the mirror made me realize why I got a strange look at the reception desk. I jumped in the shower fully clothed, shoes and all and washed off clothing and then myself. It was a memorable day.

The following day, due to the difficult ground route to the proposed test location, it was suggested that I abandon the task. This made me even more determined. With two volunteers in tow, the penetrometer, which had now arrived, and two dump trucks from a construction company, we set off. Loads dumped by the trucks made that wet/dry gulch more navigable. We waited for one of the dump trucks to be reloaded and become heavy with ballast and reached the test site. The penetrometer was attached to the chassis of the truck and the cone-shaped working end was able to penetrate the soil. A battery powered, electric motor drove a hydraulic pump, which supplied the penetrating pressure. The pressure recorded for the first penetration was encouraging and soon the runway was punctured with a series of small holes laid out in a grid. The ground was soft enough and the tests were on.

We had been chastised before for not providing heads up information as to our intentions. Years previous, after conducting gravel runway tests at a remote location known as Ed's Field, our company public relations department had to explain our presence there, to the US Drug Enforcement Agency. They thought that a large aircraft operating in a remote field might be engaged in drug smuggling. This time I did inform the local sheriff, the FBI and the DEA.

Early morning on the day the tests were due to start, I accompanied a heavy truck loaded with ballast bags and headed towards the test site. The ballast was unloaded, covered with a tarp and the truck returned to the construction company from whence it came. I had wanted the aircraft to start the tests lightly loaded, because of the unfamiliar surface hardness. With me, I had a cell phone; a novelty in those days. After a couple of hours waiting for the test aircraft to arrive, I tried to make contact with Wally Warner, our pilot at Avra Valley, with no luck. I heard later that he was desperately trying to contact me. Suddenly my phone rang. It was Joel Paul, calling from Canada to inform me that a severe dust storm, to be followed by heavy rain, was heading my way; it was approaching from the south and could be upon me soon. He had been advised to tell me to get out as soon as possible. I looked south and informed him that I could see the dust cloud rapidly approaching. Thanking him for relaying the message, I jumped in my vehicle and hightailed it out of there, hoping to get through that dry gulch before it filled with water. A few seconds later, that ominous cloud of dust reached me and suddenly I was reminded of the thick London fogs, when you could not see your hands in front of your face. I could barely see the fence beside the track and was only able to drive at a slow walking pace. It was a slow stop and start drive, to be able to follow that fence. Visibility gradually improved as the rain commenced. By the time the rain was coming down in torrents, I found myself in that gulch that had been my nemesis. It was rapidly filling up with water. Somehow, I made it through and got back safely.

We waited for the ground to become dry, then obtained several more penetrometer readings to ensure that the bearing strength of the test site had not changed. After three days we commenced testing to obtain the aircraft performance on that surface.

On each engine-cut take off we created our own local dust storm. The crude barb wire fence beside the runway and the dust reminded me of a time more than half a century ago that I stood beside a similar fence in Peshawar and got my first close encounter with an aircraft. After completing the lightweight tests, the aircraft returned to Avra Valley to fill the fuel tanks and returned in less than an hour. We loaded the aircraft with about 10,000 lbs of ballast bags, each bag a bright yellow and weighing 25 lbs. We noticed an aircraft circling high above but paid no attention to it. We were consuming our boxed lunches, supplied by the hotel, when a Blackhawk helicopter landed nearby. After the dust settled, we saw two armed men approaching us. One of the members of our team raised his video camera for a shot, but Wally Warner wisely told him to put it down, as it might be misconstrued as a weapon. Several pick-up trucks were seen nearby as the armed men from the helicopter reached us. We explained our presence there and, when I mentioned that the local Sheriff, FBI and DEA had been informed, they pointed out that communication between local and federal departments was virtually non-existent. After a quick examination of the contents of some of the ballast bags and wishing us good luck with our tests, they left in a thick cloud of dust. Apparently; a spy satellite had spotted our operation, a spotter plane had been dispatched followed by the helicopter and agents were sent as a back up in pick up trucks. If we were genuine smugglers we would not have stood a chance against capture. It was a most impressive demonstration of modern technology in action. By the time the misunderstanding and the disruption the tests was over, it was too late to complete the tests that day.

We completed the tests the following day. Before leaving, I went with Wally to the building site where the frame of a small hangar was being assembled. Before entering the gate of the compound, we were waved off and directed to a small gap in the fenced area. Once in, we thanked the owner for the use of the airstrip and noticed a man carrying a shotgun. On asking him what he was hunting for, he simply replied, Indians and promptly placed his gun in a small sentry box. He then explained that he was on sentry duty that night, to protect against Indian raids. During recent nights, trucks had been fired at and radiators were damaged, also equipment, even including firewood was stolen. It was indeed a feud worthy of a Zane Grey Western novel. The main entrance we were shepherded away from, had been booby trapped with the sharp ends of large nails sticking out of planks, lightly camouflaged with desert sand.

I returned the following day and faced that dry gulch but not alone. This time I had members of the team with me. We brought gifts for the owner, drill digger, and his squaw. The men received Dash 8 baseball caps and belt buckles and posters of the aircraft, the lady was given a company scarf, also a belt buckle and baskets of fruit and flowers. The gift most appreciated was, a box full of lighters that would eliminate the necessity of short-circuiting a battery to light up a cigarette.

It was getting close to Christmas. The team had missed the company Christmas party, so I thought it was time to throw them a party and invite the folks at Avra Valley and the desert strip crew we got to know.

A restaurant in Tucson that boasted a belly dancer and a meal consisting of a complete roast lamb on a spit, with all the and trimmings, was chosen. Invitations were given, mostly by phone and soon it was party time. The word had got around and there were many more that attended than I had estimated. The restaurant owner, produced more tables and food and a good time, was had by all. The belly dancer that was an excellent example of a healthy beautiful female and tried to coax volunteers to dance with her. Only two responded; I was one of them. We both received lots of cheers, or were they jeers, for our futile efforts. Towards the end of the party, Wally and I both thanked everyone for the support we received. The manager of the Tucson Aero Service Center spoke of the dogged effort made to complete the operation out in the desert. For my desert experience, I was made an honorary cowboy and a genuine Stetson was placed on my head. It was a gift I was proud of and I wore it all the way home, arriving there on Christmas Eve.

It was near the end of January, relaxing in front of a wood fire in our winter wonderland in Muskoka, when I received a call from Bombardier. It was Joel once again asking me if I would participate in a test program. This time it was not to be in sunny warm Arizona but to cold

North Bay in Northern Ontario, almost within commuting distance from my home. Bombardier were busy with the Dash 8 Series 400 tests and were unable to participate in the tests undertaken by a joint NASA, FAA and Transport Canada, test program. They were able to provide a Dash 8 aircraft piloted by Barry Hubbard and required me to co ordinate the tests with a joint team headed by NASA. I accepted the task and drove the one hundred kilometres to North Bay to get a briefing on the program, which was well underway. The Dash 8 was the last type of aircraft to enter the program, which had commenced the previous winter.

The task was to measure braking capabilities of a variety of aircraft under adverse runway conditions, such as snow, slush, and a variety of icing conditions, including, the effect of sanding down an icy runway. In conjunction with these tests, ground vehicles from many countries, used to report runway-braking friction to the control tower, would run down the test runway and provide data prior to each breaking test. Many countries had their own standards when reporting braking conditions. The goal of this program was to apply a universal ground-vehicle braking standard, and determine how that standard affected different types of aircraft. Jet, turbo prop, high wing, low wing, big and small aircraft each provided different braking capability on the same breaking surface. In the future, an aircraft landing on a contaminated runway, anywhere in the world, would be aware of the distance required, based on the data supplied by ground friction measuring vehicles.

The Dash 8 aircraft was equipped with a satellite antenna and a Ground Position Satellite indicator, which was added to the existing instrumentation. In order to gain the required accuracy we placed an additional satellite transmitter on a tripod located on a carefully surveyed spot near the runway. Now the position of the aircraft could be recorded with an accuracy measured to an inch. Accurate ground speed, braking friction and rolling friction could now be obtained. When conditions were right, on the surface of the test runway, a phone call to Barry Hubbard at Downs view, quickly produced the aircraft rapidly at the test site. This caused minimum impact on the high priority Dash 8 Series 400 test program at Downsview. In this manner, two depths of a snow-covered runway, an icy runway, and an icy runway, sprinkled with sand, were conditions that were tested. The latter condition reminded me of days gone by when the railway tracks on a mountain incline in India, were sanded to improve friction.

Each series of tests commenced with ground vehicles from Norway,

UK, Canada, USA and a NASA vehicle, all recording breaking friction to their own standards. The aircraft then entered the test location at a variety of speeds, using brakes only as well as brakes combined with the use of reversing propellers. Tests without the use of brakes or reversing propellers were also undertaken to obtain rolling friction. Barry had recently transferred from a test engineer to a test pilot and I was impressed by his calm and professional approach to the task. It was a pleasure working with him again.

I hoped my small participation in those joint tests would lead to an improvement in aircraft safety when operating under adverse contaminated runway conditions.

CHAPTER 26

It was a sad day in 1997 when I received news that Jock Aitken had suffered a heart attack and died. He was due to join Stuart Nicholson and myself in retirement, but that was not to be. He was the first of the group of three that joined the de Havilland Flight Test Department from Shorts in Northern Ireland. In his will, Jock expressed the wish that his ashes were to be scattered over the area, which most test flights from Downsview took place. Wally Warner who was once Chief Pilot for the Ontario Department of Natural Resources and is currently the Chief Test Pilot for the de Havilland Division of Bombardier, used his influence and obtained the use of a single engine Otter aircraft from his previous employer. The aircraft had been used for seeding the lakes with fish. A cone-shaped hopper installed in the center fuselage, had been used to drop small fish or fingerlings, to maintain or increase the fish stocks, in the many lakes in Ontario. That hopper was ideal for fulfilling Jocks request.

Many years previous, a pilot expressed the wish that, after his death, his ashes to be scattered over the Atlantic. A Beaver aircraft on a ferry flight across the Atlantic, attempted to drop his ashes, but had to return to Halifax to vacuum up his ashes, which had blown back and were scattered on the cockpit floor. The second attempt was successful, when a hollow tube was stuck out of the side window and after a successful test run with sand, his ashes were deposited.

Jock's wife Jean, placed Jock's ashes in a decorative cardboard box, Stuart and myself waited for Wally and the Otter to arrive at the Muskoka airport. After the Otter arrived, we all jumped in and headed out looking for my home and property, which was in the middle of the test area. I remember being out there on holiday with a communication radio, listening and watching a Dash 7 conducting stalls and engine handling tests directly overhead. Jock was on board above me, during the engine handling exercise, so when we reached my home I knew his ashes would reach his intended destination. Stuart had a firm grip on the box when he handed it to me. I removed the plastic bag containing

Jock's ashes and just as I reached the hopper, the bag burst open at the bottom, and his ashes dropped into the edge of the hopper and were spread according to his wishes as we circled the area. I could just picture Jock writing a snag sheet and sending out one of his queries to engineering design. He would snag the bag for inadequate containment, call for a material inspection, a stress analysis and a design review. Jock would always make sure that a defect was thoroughly dealt with. His expertise and sense of humor will be missed. May he rest in peace! The scattering of his ashes caused me to recall the many funeral pyres in the Hindu holy city, Benaris, now Vanarisi India, where the ashes of loved ones were scattered into the holy Ganges River.

Back in retirement, life was good and never boring. The day I retired in 1991, I vowed never to carry a watch, and never did. As with many retired couples, we travelled, taking road trips to Arizona, Florida, and British Columbia. Air trips were made to England, Scotland, Ireland, France and Spain, to visit friends and relations. My siblings had spread all over the globe. Cecily was in the South of France, Peter was in Spain, Winston was in Borneo, Margaret was hard to track. We visited her in San Francisco, Scotland, Jamaica and Tobago. She also lived in the Hawaiian Islands, Tahiti, Thailand and currently resides in Malaysia. David lives in England and has a holiday home in Spain.

The word cosmopolitan suits our family. We had booked a trip to visit Australia to both tour and visit my many cousins who I had last seen when we marched with the Afridi tribesmen in Peshawar, India, a long time ago. The trip was also to include New Zealand, where I had always wished to visit.

It was just after New Years Day in 1998, that I received a call from a consultant company, asking me to participate in certification tests of a Penn Turbo Caribou aircraft. It was a standard de Havilland Caribou, which had its piston engines replaced with turboprops. After an interview over the phone, I was asked to go to an airfield near Cape May, New Jersey, where I would be briefed on the task. This was the second time that the Consultant Company had contacted me. Several years previously, I had accepted a contract to participate in certification testing on the aircraft. Just before I was due to join the program in New Jersey, I heard the dreadful news; the aircraft had crashed on take off, en route to New Jersey, killing all crew members. I had dodged that bullet and once again, the Gods were with me. The pilots on board were not test pilots and had limited experience. When the program restarted, with a second modified aircraft, an experienced and well-qualified team had been contracted to undertake the certification tests. As usual with development tests prior to certification, many problems and setbacks occurred. Program delays, when combined with cost overruns, created a conflict between the company management and the test team. This resulted in the test team abandoning the task, leaving the program high and dry. This was the situation when I arrived on the scene. The good news was that Stan Kereluk had been hired as test pilot. He was an experienced test pilot who had recently retired from test flying with the National Research Council (NRC), based in Ottawa. Stan had flown the Augmentor Wing Buffalo Research program and had conducted many tests associated with the stability augmentation system. I had got to know Stan during several visits to NRC in Ottawa.

Stan and I spent a week studying test data and reports generated by the previous test team. We came up with a test plan to complete outstanding tests. There was a significant amount of analysis of the aircraft performance data and more handling and performance tests, to be undertaken. I realized that we needed help and contacted Stuart Nicholson to check on his willingness to participate. He had retired as Chief Performance Engineer with Bombardier; he was the perfect candidate and was also willing to help. After a quick visit to the consultant company, he joined us. It was a small three-man team of retirees, well experienced, whose personalities blended well together. We were staying in a small motel, in a small town close to the test airfield. We were given the use of a rusty old car, which managed to take us to the airfield and back every day. The Penn Turbo Company manager took us out to breakfast, lunch, and dinner and paid for the low budget motel, so we had no expenses. It was a low cost operation with a small team.

My location in the aircraft during tests, was mid fuselage, where the computer recorded data and also provided me with data visibility. My parachute exit, in an emergency, was the rear ramp door. It was a similar setup as that on the Buffalo aircraft I had flown in many years ago. Experience had taught me that a rope with good knotted handgrips, leading to the exit in case of an upset, would be invaluable. I was now in my seventieth year and wondered if I would be nimble enough to climb up that rope, if the aircraft was in a nose down position, to be able to reach the exit. I believed adrenalin would get me to that exit.

After five weeks of intensive flying, it came time to demonstrate flying qualities to the Airworthiness Authority acceptance team. Stan had gone home to Ottawa for a few days on personal business. He owned a hobby farm on the outskirts of Ottawa and had a herd of Beaffalo (a cross between domestic cattle and buffalo). A problem with the herd was too difficult for his wife to handle alone.

The Transport Canada pilot was comfortable flying with the Penn Turbo experienced co-pilot, so we commenced the acceptance flights. We were asked to include a demonstration of rapid entry stalls. These stalls were not conducted on the aircraft during its original certification with piston engines, as rapid entry stalls were not a requirement at the time of certification. Failing in my argument that new regulations should not apply to an old airplane, I reluctantly accepted that rapid entry stalls were to be conducted.

The certifying authority test pilot was a stout fellow and with plenty of girth around his waist. On the day of the rapid entry stall test, the weather was marginal. I was sitting at a test instrumentation console located mid way down the fuselage. Knowing the history of previous stall tests conducted on the Caribou, I felt quite nervous and made sure that the knotted rope between my seat and the rear exit was securely fastened. Before the test commenced, I went forward to the cockpit and commented that visibility was poor. I also realized that any experienced test pilot that I had previously flown with, would have called off the test. However, the test proceeded not withstanding my objections. A not-so-rapid entry into the stall was conducted and several stalls had to be repeated to obtain the required entry rate. Post flight analysis of the flight data, revealed that the required rate of entry was not achieved. Why? Because full up-elevator was not always achieved, due I believe to the pilot's belly preventing the control wheel from rapidly going all the way back to the stop. When Stan got back, all went well and the stall was successfully demonstrated.

As previously agreed, I left the test program a couple of days before my holiday trip to Australia and New Zealand. There was not much flying left to complete the program when my replacement sent by the consultant company, arrived. Stan and Stuart carried on with the remaining test program,;the new guy was a disappointment, lacking experience, so I was told. For some reason, it took a lot longer to obtain certification than I had envisaged. Test flying was the easy part. There were a lot of i's to dot and t's to cross, with respect to the drawing changes and modification standards that would clearly define the aircraft that was to be certified. This kept the consultant company busy for quite a while, until a Certificate of Airworthiness was issued.

We were back home from our most enjoyable six-week trip, having driven 10,000 miles in Australia and 5000 miles in New Zealand.

Summer was nigh, when I received a call yet again from Bombardier. It was Joel asking me to build a long water trough and coordinate water contamination tests on a Dash 8 Series 400 aircraft. Tests were being conducted in Chippewa, where I had previously built a 250 feet long water trough. This time the trough was to be a lot longer as they wished to include measurements of wheel drag. It would be quite a challenge to find a suitable area to meet the tight tolerance for the water depth required. I accepted the challenge and was soon on my way to Chippewa, Michigan.

I was given two men to assist with the survey and build-up of the trough, which had to contain water to the depth of half an inch with a tolerance of plus or minus a quarter inch. The laser transit sight I rented was not accurate enough for the task. Repeat surveys gave inconsistent results. A different technique was required. After scratching my head, hoping for a light bulb to glow, I went to a local hardware store and purchased two metal yard sticks, 100 feet of transparent water hose and two, three-foot lengths of lumber. The tube was filled with water, colored with food coluring, obtained from a grocery store. Open ends of the tubes were attached to the top of the yardsticks, which in turn were fastened to the lumber. Now, we had a long U tube or spirit level, which gave an accurate and repeatable reading of the height difference between the yardsticks. After two days of a painstaking survey, a 700 foot length was chosen on a long taxiway. The sides of the trough were built using small lengths of closed cell Styrofoam, glued to the pavement with liquid nail material, which was fast drying. The two ends of the trough and the two dams required to maintain the half-inch of water were made of rubber strips with rubber ramps for the aircraft wheels to roll over. Spill gates were installed at strategic locations to maintain water height in case of overfill or increase due to rain. When the trough was filled and inspected by the Transport Canada test witness, he reported that the long trough was an engineering marvel. Little did he know that luck in finding the right location played a big part. Two days were spent with the Dash 8 series 400, dashing through that gigantic puddle using a variety of speeds and flap settings. The spectacular splash created after each run, required the trough to be topped up, using the water hose on the standby fire truck. Water did engulf the engine air intake, but the auto ignition maintained the jet flame without loss of engine power.

While I was in Chippewa, I was able to renew friendships with many old colleagues including Joel Paul and Wally Warner. It was a welcome break in my retirement and I thought, my final ride in the test saddle. That was not yet to be.

My final test participation was in October 1999, eight years after retirement. It started with a phone call from the Bombardier Product Support Department, asking me to help Trans Canada Airways to obtain narrow runway certification for the De Havilland Dash 7 aircraft. They were the first operators of the Dash 7 and required to operate on a 14-metre wide runway in Guiana, South America. Bombardier, apparently, were unable to provide support. If they were to undertake the task, with the then present management, it would have been at considerable cost. The Project Office, Airworthiness Department, Technical Design Department, Estimating Department and the Flight Test Department, including Instrumentation Design, would all be involved. With overhead costs and numerous meetings with a multitude of attendees, the price tag to a small company like Trans Canada Airways would have been out of reach.

Having completed narrow runway certification on both the Series 100 and 300 Dash 8 aircraft, during my earlier days in retirement, it appeared a simple task and I agreed to form a team to allow narrow runway operation for the Dash 7 operator. Stuart Nicholson once again offered to help. He would alter the Flight Manual to reflect the changes due to the increase in take off speeds and also work with me on data reduction and analysis. Stuart would also coordinate the tests by radio at a portable wind station to be placed beside the runway. The third member of the team was King Chiu, who had previously helped me with my remote sensing mail alert. King, having headed the de Havilland Instrumentation Department, would be invaluable in designing and installing the necessary instrument recording package, as well as any special provisions for the tests.

A visit to Trans Canada Airways, located at the Toronto Island airport made them aware of the task in hand. They were advised that an experimental flight permit was required. The modifications to be installed were, test instrumentation, an engine fuel cut off, a high pressure valve located at the fuel control unit, a co-pilot operated cut switch, and a water drip kit to measure runway deviations. This necessitated the aircraft to be taken out of service until the modifications for tests were removed, allowing the operating permit to be re-instated. The test runway had to be painted to represent the 14 metre narrow runway. Since the Island Airport was short and very active, mostly with light aircraft, I considered it unsuitable for the tests. It was hard to believe, but a serious suggestion was made to conduct the tests on the actual narrow runway in Guiana. Had we foolishly agreed to this, the aircraft would have been damaged when it ran off that runway during the necessary development phase of the tests. I did suggest that it was most desirable, if not essential, that the development phase of the tests be conducted by an experienced test pilot. All except the installation of an engine-cut test system was reluctantly agreed to. It took a lot of explaining, to get agreement for the incorporation of the test engine-cut switch. It may be of interest to the reader to explain the reason. There have been many engine-cut take offs as previously mentioned and it is time for an explanation of the cut process.

During each take off, when all four power levers or throttles of the four-engine Dash 7 are moved forward by the pilot, switches are triggered to arm the auto feather system. In the event of a sudden engine failure after, the decision speed is passed and after a two second delay, the auto feather system, sensing the loss in power, will automatically feather the propeller on the failed engine. Feathering the propeller, places it in a low drag position, thus improving the aircraft climb away performance. In the event of engine failure before decision speed is reached, the pilot will throttle back all power levers and automatically cancel the auto feather system and bring the aircraft to a halt. Without auto feather, the propeller on the failed engine will windmill, creating high drag and assisting the aircraft to come to a stop. Simulating an engine failure by the pilot simply throttling back the critical left outboard engine to idle, was suggested. This was not a suitable representation of a engine failure on takeoff. It would not only give an unrealistic results with respect to runway deviations, but would also pose a hazard, as the auto feather system would be disarmed, preventing automatic feather action and leaving the propeller in a high drag condition during a critical period during a continued take off. The other suggestion made was to use the firewall fuel shut off switch to cut the engine. This fuel shut off switch is to be used in case of an engine fire and would prevent fuel from feeding the fire. Using this switch would delay the engine failure while fuel remaining in the long length of pipe, still fed the engine for a while. In addition, using the firewall switch with the engine at high power could cause the fuel line to collapse or be damaged due to sudden low pressure with the potential of causing a serious fire hazard. *Little knowledge is a dangerous thing.*

Once the go ahead was given, a compliance test plan and details of the tests were sent to Transport Canada. Our three-man team installed the modifications in only two days. King Chiu's knowledge and experience was invaluable. He purchased for his own later use, a signal control box and connected it and his lap top computer to the aircraft crash recorder. The Crash recorder provided all the necessary instrumentation required for the tests. Instrument calibrations were completed and with the computer equipment strapped to a passenger seat and the water drip tap close by, we were ready for the tests.

I set off to Hamilton airport, the agreed site for the tests, to check the runway marking, simulating a narrow runway. The airport manager was made acquainted with the test details and then he dropped a bombshell. He would not allow spotters on the side of the runway to continuously run out on to the runway and measure aircraft deviations using the tell-tale water trail. My statement that spotters had been used on many other airfields, fell on deaf ears. I had to come up with plan B.

When originally contemplating methods for measuring runway deviations, I had considered using a video camera. Now was my opportunity. A rented video camera with a wide-angle lens was strapped to the rear tail bumper of the Dash 7 using heavy duty tie wraps. The camera was connected to a video recorder and monitor that could be viewed and operated from my seat location. A check in the hanger, with a chalk grid marked on the floor gave confidence that accurate measurements could be obtained to measure deviations from the centerline of a runway. The video camera gave almost a worm's eye view of the wheels and runway.

The next hurdle was that an experienced test pilot was not available. The company chief pilot, R. McDorman, offered to conduct the tests. I went through a detailed review of the tests. He was surprised that the use of nose wheel steering was not allowed to control direction on the ground. A preliminary flight was conducted at the Island airport to check out the test equipment and also served to give me confidence in the pilot's attitude and capability. After using the engine cut switch during several take offs at reasonably safe speeds, I was satisfied to continue with the tests.

At Hamilton airport, as the tests progressed, I became aware that my task in the aircraft was pretty hectic. The task was, logging initial deviations off the video monitor, checking the computer for airspeed at the engine cut, logging test time, aircraft test configuration and test weight, communicating with the cockpit and wind station, saving test data on the computer, and directing the tests. When I missed saving a test run on the computer, King Chiu offered to come aboard and work his computer. When at de Havilland, King had expressed his reluctance for participating in test flights. He deserves a lot of credit for helping me on board for the tests. He was calm, cool and collected as he sat next to me while the engine was cut and the aircraft lurched to the left on each and every take off.

We completed the tests and established the required take off engine cut speed for the narrow runway. The three of us were jokingly referred to as the geriatrics test team.

It was now time for the Transport Canada pilot and test engineer to verify our results. All went well as we demonstrated aborted engine cut take offs, and landings, including crosswind landings on the simulated narrow runway. When we got to the continued take offs with the large take off flap setting, runway deviations were greater than we had ever seen, well beyond the side boundary of the narrow runway. Increasing the cut speed made matters worse, not better, as it should have done. The company pilot took over to repeat the test and he also could not contain the aircraft, anywhere close to the side boundary. It was an embarrassment and extremely puzzling and time to call a halt and try solve the puzzle.

After examining the video, showing deviations and the computer data and checking the aircraft airspeed system, there were no answers late that night. I must have been working that problem in my sleep and woke up suddenly with what I hoped was the answer to the problem. Early next morning, careful re-examination of the video, confirmed my thinking, which was also supported by a ground crew member's description of the incident. The large deviation cases recorded showed the main wheels were off the ground at the time the engine was cut, with the nose wheels still on the ground. We were akin to an unstable wheelbarrow. The main wheels were unable to provide the restraining side force to reduce the deviation. The Dash 7 was crying out that it had no business being on the ground at lightweight and at a speed that wanted it to get off the ground. The pilots were pushing the control wheel forward to keep it on the ground. Understanding the problem was one thing; finding a fix was another.

Two variables were examined during the next few flights. The pilot's wheel position at the time of the cut and the aircraft weight. The wheel position did affect the deviations but since it was a variable that could not be adequately controlled, it was of interest, but of no value. Increasing weight above the test weight that showed the deviation problem, eliminated the problem. A minimum weight could be controlled, was acceptable to Trans Canada Airways, and was a limitation that Transport Canada could apply to the aircraft type-approval for operation on 14-metre narrow runways. Problem solved. It took 15 flights and 7 days to complete all tests. It took Stuart and myself three days to submit a test report to Transport Canada. The report included proposed changes to a supplemental flight manual to reflect the performance changes due to the necessary increase in takeoff speed. Approval to operate on 14-metre wide runways was granted a few days after receipt of the test report. It was not a bad effort for a small geriatrics team.

Logical thinking sometimes reaps great rewards. I had come to an age such that the feeling of immortality in my earlier years, was no longer present. I thought that waiting to leave an inheritance ,when I was no more, did not make sense to me. A partial advance inheritance, one that we all could enjoy made a lot of sense, so I put a large down payment on a log cottage on the South branch of the Muskoka river, about forty kilometres from my home. My three daughters would jointly own the place and pay the remaining costs. It was at that beautiful spot, that I took a boat, with an outboard engine to explore the river. The boat was steered by moving a tiller arm attached to the motor. The awkward body twisting, while looking forward and steering made steering tiresome, causing back and shoulder pains. There had to be a better way.

I believed that my flight testing days were over. To keep my development and test juices flowing, I came up with a fun project, which could provide easy and relaxing control for small boats.

The first approach made to solve the problem was to steer the boat with a two-way wander switch.. A battery supplied the power to an electric motor. The motor moved a rod, which was connected to the tiller by a quick-release mechanism. It was a similar set up to the one used on the previous remote controlled lawn mower project and just as responsive. While attempting to go through the wake of a passing boat at high speed to check ability to steer in rough water, the quick tiller release, released too easily and control was lost, due to buffeting in the waves. Before I could regain control, the boat struck the river bank and propelled me to the front of the boat, leaving me bruised and embarrassed by my over sensitive quick-release design. A short while later, when sitting in the swivel seat which was recently installed, I pondered back in time and remembered sitting in the tail gun turret of a Shackleton aircraft, where the seat could be moved up, down, left or right and was linked to control the pointing of a gun barrel. That light bulb switched on once again; why not let the seat control the tiller? it would allow hands-free steering. It was back to the drawing board to come up with a design that eventually led to obtaining Canadian and US patent for Body Use Motion (BUM) Hands Free Steering for outboard motors on small boats, referred to as BUM Steering.

There were a few set backs during the early design phase, the most embarrassing set back came when demonstrating the BUM Steer at a boat show.

A local TV station asked for a demonstration, which was to be aired on the news program, which was due to start in five minutes. Excited by having my project given media coverage, I jumped in my boat and with cameras rolling proceeded to demonstrate the instinctive and relaxed steering control. Both hands up in the air, swiveling and turning to the left and right and even performing a tight pirouette, the watching crowd applauded; leaving me proud of my invention. My pride soon turned to dismay when I demonstrated a new feature I had installed the day previous to the show. This new feature that I demonstrated, allowed me to disengage the seat from the motor tiller and re-engage the seat in a side-facing position. This would allow a flexible seat position for controlling direction while fishing off the side of the boat. An incorrect engagement led to a control reversal, such that, when I swivelled to the left the boat went right. I disengaged and reverted to standard tiller operation, it was too late and I banged into a large expensive boat moored near by, causing a small paint scrape. All this was featured in the evening news and seen by friends and old colleagues, adding to my embarrassment. A re-design was in order.

It was snowing on the river and a race against time, before the river froze over, when the Bum Steer passed the function and reliability tests set to meet my aeronautical background test standards. I applied for a patent and with the help of a previous co-orker, Peter Piri, whose computer drafting skills far outweighed my crude sketches, I received it. My next project was this book and when it is completed, I will be searching for another project to keep the brain active. Perhaps it will be a fiction novel.

CHAPTER 27
Generic Tests

This chapter puts together many essential flight test tasks that are generic and apply to all aircraft. It is included to bring confidence to the air traveler by providing an added insight into the world of aviation and the effort that goes into taming aircraft to carry precious loads.

SPEED DETERMINATION

Air speed errors, Take off speed, Landing speed and Max Operating speed determination, does not sound too complex a subject, but obtaining these speeds can be both hazardous and complex. They however provide adequate safety margins for commercial air operations.

AIR SPEED ERROR TESTS.

Air is not the same at different altitudes and temperatures. Air is denser at sea level than it is at higher altitudes. At extreme high altitudes there is hardly any air. Temperature also has an effect on air density, which decreases with high temperatures and increases with cold temperatures. For aircrew to know the speed over the ground under zero wind conditions, accurate altitude and temperature must be obtained.

Corrections for wind conditions are also required. An accurate measurement of air speed and altitude is necessary for safe flight.

Air speed error measurements are usually conducted in the early days of a test program. Several aircraft have crashed conducting speed error tests. Almost the whole Handley Page Aircraft Company test team was lost, when the Victor aircraft crashed while conducting low level (50 feeet) Speed-error tests. An unproven aircraft flying perilously close to the ground was really an accident waiting to happen, and it did.

In order to measure airspeed on test aircraft, a test air data boom, which can be about 10 to 15 feet long, is mounted, usually near a wing tip, or protruding from the nose. This boom contains two high-speed wind vanes. One, a pitch vane, measures the forward angle of the airflow and the other, a yaw vane, measures the lateral direction of air flow.

Also mounted on this boom, is a probe, called a pitot-static head. On test aircraft, this probe is designed to swivel, and always points into the wind. This probe measures, both the pitot pressure, akin to the pressure you feel pushing your hand back when you stick it out of a car window and atmospheric or static pressure, the pressure as measured by an altitude indicator or barometer. Both pressures are sensed by indicators, which present airspeed and altitude in the cockpit.

The air speeds measured by this boom, are reasonably accurate but the accuracy of the test system, as well as the basic aircraft air speed system, must be within an acceptable tight accuracy tolerance, defined by Airworthiness requirements, hence errors must be known and carefully measured.

The test data boom is not suitable for use on a production aircraft. On these aircraft, the static pressure is sensed by a static vent plates located on both sides of the fuselage, near the cockpit area. Dynamic (pitot) pressure is sensed by fixed pitot heads, which are also usually located both sides of the fuselage. Correcting this for errors is easily done, by comparing the pressure from the test swiveling pitot, to the fixed production pitot on the fuselage.

Correcting for static pressure errors is not as easy. Local airflow over the static vents cause errors in the recording of static air pressure. This creates errors in both air speed and altitude.

The first static position error test (P.E. tests) I took part in was at the Aircraft and Armament Experimental Test Establishment at RAF Boscombe Down near Salisbury in the U.K. The test aircraft was the Avro Shackleton, an RAF costal command reconnaissance four engine aircraft.

Two sensitive pressure-measuring devices, known as aneroids, were used. These could measure pressure height accurately, within inches. They were placed beside each other in the control tower and were set to read the same. One aneroid was taken to the Shackleton aircraft and connected to record the static vent pressure. The other went to the reference tower.

A camera ,with a grid in front, was located beside the aneroid in the tower. Due to the high sensitivity of the aneroid on board, the aircraft was restricted to a couple of hundred feet above ground. The task was to fly at a steady speed 50 ft above ground, and when over a reference point, record both tower and aircraft aneroids and photograph the aircraft at the same time. When the static vent on the aircraft was at the same height as a clump of trees, as seen on the developed photograph, any difference in aneroid readings would be due to static position errors.

Knowing heights above or below the clump of trees would allow correction for test errors. Many tests were conducted at a variety of speeds and flap settings. To ensure steady state speeds, all test points had to be conducted at very low levels in a racetrack type circuit, requiring extreme concentration on the part of the pilot. It is not surprising that several crashes have occurred using this test method as used at Boscome Down.

Safer methods for conducting these tests were later devised. One method was to use a pacer aircraft, which allowed tests to be conducted at a safe altitude. The Royal Aircraft Establishment, located in Farnborough in the U.K., supplied the calibrated pacer aircraft. With the airspeed system carefully calibrated, these aircraft were offered to the aircraft industry for air speed calibration use. The Gloster Meteor, a twin jet fighter aircraft, and later, the Gloster Javelin, a delta wing jet fighter, was made available for air speed error tests.

My experience using the pacer aircraft was on the A.V.Roe Delta wing Vulcan Bomber and the English Electric high altitude reconnaissance PR 9 Canberra aircraft.

Not all pilots including test pilots are experienced in close formation flying. The Vulcan tests, with Rolly Falk as pilot went without incident.

Each test speed point was synchronized with a recording on the Meteor aircraft, which also took a photograph to identify any height variation between us. This height determined the static position error.

The Canberra PR 9 we used for the test was not equipped with data recording. My location in the aircraft did not allow viewing the cockpit indicators. The pilot, Dicky Turley- George was required to read six items from his cockpit instruments, which I would write down. This distracted him from maintaining close formation. After our wing tips made contact, resulting in wing wobbling, I politely suggested that such close formation was not necessary. This *"safe"* method could have been catastrophic.

The Canberra PR 9 also calibrated by the aneroid fly-by method, using the top of the famous Blackpool tower. Being a high tower, it provided some safety to the operation of the tests. A two-man team was sent up the tower to photograph the aircraft and record air pressure for each run.

There was a bit of a delay in starting the test, due to the time taken to take one of the aneroids from the tower to the aircraft. The tests would provide great entertainment for the large crowds always on that stony beach in front of the tower. We were ready to start our first run, which would be three hundred feet high over the sea and a few hundred feet off the waters edge. The tests had to be abandoned, as there was no radio communication with the person in the tower. It turned out that person with the radio had gone down to the large bar located at the base of the tower. He happened to be an alcoholic, so that ended that day's operation.

A few days later, with a sober ground crew, the tests were successfully completed much to the interest of the beach crowd.

A short time later, a similar crowd would witness, a spectacular crash, when the same aircraft flying at low level, shed a complete left wing. This incident has been described earlier.

A trailing-cone method was also used for position error tests. Developed in a wind tunnel, a fiberglass cone towed sufficiently behind an aircraft, gave accurate pressure readings. This method is currently used on many prototype aircraft.

When attempting to use a trailing cone on the de Havilland Buffalo military transport, due to the shape of the rear fuselage, the trailing cone was very unsteady and gyrated quite violently in all directions, causing this method to be abandoned.

A speed course method was used on the Buffalo. A ready-made surveyed speed course was available near by. The Ontario Provincial Police had placed markings on a long straight section of highway 400 near Toronto. These markings were used, in conjunction with a police spotter aircraft, to identify speed violators by timing cars between markers.

Under calm air conditions each test speed was conducted in opposite directions to counteract the effect of any wind. Tests were flown at a reasonably safe height, which would still allow accurate timing between markers. This height was targeted to be 400 ft above the highway.

All required test runs (approximately 50) were conducted. After returning to the airfield, we were informed that the company was inundated with phone calls from citizens, police, and radio and television stations. Some calls were complaints about low-flying aircraft, some about noise and some about an aircraft in difficulty, trying to land on the highway. Note, some tests were conducted with landing flaps and wheels down.

All the speed position error tests described so far covered a normal speed range. A low speed range cannot be investigated by any of the methods described because of the hazard of flying at speeds near the stall, which is the speed below which the aircraft ceases to fly and literally falls out of the sky.

To measure airspeed position errors at or near the stall, a bomb shaped object, with fins at the back and a probe in the nose, is towed below and behind the aircraft. This probe has holes in its side to measure static air pressure, This test cannot commence until the aircraft has undertaken stall tests to ensure safe stalling characteristics. A bomb attached to a cable, wrapping itself around the aircraft is not a good idea.

After stall-handling tests are complete, the trailing bomb can be installed and stall speed error tests can begin.

Stall tests

An aircraft stalls when the airspeed is too slow to sustain flight and the aircraft literally falls out of the sky. Safe recovery from a stall must be demonstrated in order to certificate an aircraft as airworthy. The stall speed must also be determined to make it possible to provide adequate safety margins and to prevent the aircraft from inadvertently stalling in service.

To meet stall-handling requirements, as with all stability and control tests, the aircraft must be examined under a full range of weight, center of gravity and engine power settings.

In order to investigate problems arising from stall tests, wool tufts are often used to visualize the airflow just prior to, and through the stall. Hundreds of wool tufts approximately four inches long are taped on the wing and tail surfaces, or on any area that has developed suspicious air flow characteristics. The wool chosen is a color that contrasts with the surface to which it is attached.

A chase plane capable of flying at a very low speed can fly alongside the test aircraft. On board the chase plane, typically, there might be a cameraman with a high speed or a video camera and a few engineers, all eagerly studying the wool tufts, as the flow pattern changes. Beginning from a steady state flow, prior to the onset of the stall with tufts aligned to the airflow, the speed is then decreased and the tufts commence to wiggle and flicker increasingly as the speed drops to the stall. Some tufts might even reverse and point forward while others would be flickering in every direction.

Notes taken by the observers in the chase plane and the play back of the camera footage, can later lead to modifications. These could be the installation of a stall bar placed strategically on the leading edge of the wing. The stall bar is typically a twelve inch long piece of aluminum, bent into a one inch triangular section with the pointed edge of the section facing forward into the airflow. In preliminary temporary efforts to find the best location on the wing leading edge, sometimes a wooden bar of triangular section is taped to the leading edge, is finally fixed, a aluminum stall bar riveted in place, will become a permanent fix.

Another remedy for unacceptable stalling characteristics is a wing fence. This is made of aluminum approximately five inches high, aligned fore and aft on the top surface of the wing between the leading edge and trailing edge. The purpose of fence is to delay or eliminate the spanwise progression of the turbulent airflow in a developing stall. Again, the position chosen for the fence might be disclosed by viewing and photographing the movements of wool tufts, fitted to the upper surface of the wing, from the chase aircraft. Several flights are typically required to finalize a fence position.

Vortex generators(small triangular or rectangular fins), are strategically placed, to encourage the airflow to remain attached, delaying the stall. Several flights are also required to finalize their locations.

If these quick remedies fail to provide acceptable stalls, more radical and expensive modifications may be required, ranging from wing leading edge changes to even major changes in wing shape.

After satisfactory stalls have been achieved, if the natural pre-stall buffet is not found to meet the certification requirements and does not give sufficient warning, artificial stall warning systems are incorporated.

Stall warning systems are typically designed to be felt by the pilot as a fore and aft shaking of his control column or wheel. Warning system must be developed, tested, and installed to give adequate warning prior to a stall occurring. This warning comes in the form of a stick shaker, triggered by a heated vane, mounted near the leading edge of the wing. When the warning is set correctly, the pilot will feel the controls shaking and sometimes rattling in the palm of his hands. This warning gives him adequate time to take recovery action.

Modifications made in response to stall problems, can cause many tests to be repeated over and over. It is not unusual for over a thousand stalls to be conducted before a stall test program is completed.

Stall tests are conducted at heavy-weight and light-weight loadings and at a maximum forward and a maximum aft center of gravity. A variety of engine power settings are to be evaluated, including idle engine power and maximum engine power levels, asymmetric power, with the critical engine shut down. Dynamic stalls (rapid entry into the stall by pulling the nose up rapidly into the stall), must also be demonstrated.

Tests are required to show the ability to safely recover from the stall and that adequate warning, prior to an aircraft stalling, is provided.

An acceptable stall occurs when a wing does not drop more than twenty degrees before a nose down pitch occurs while the pilot is still pulling the elevator control aft to raise the nose. The nose-down pitch, occurring in spite of the opposite action of the pilot, allows speed to increase to aid recovery.

Handling stall tests can be quite hazardous and must be conducted at a safe altitude and on a day when the horizon is clearly visible to the pilot.

Unsafe recovery from stalls has led to the loss of many aircraft, especially those designed with high T-tails. (A tai plane and elevators sitting high atop the tail fin and rudder) and or aft mounted engines.

In the early 1960's a number of accidents occurred on T-tail aircraft. This was due to the airflow over the elevator being blanketed by the fuselage and/or aft mounted engines, in a stall. Either of which can prevent airflow from reaching the horizontal tail. This can render elevator controls useless for recovery.

Many modern transports with this tail configuration are now designed, not to be stalled, intentionally or otherwise. They require a stall "barrier" or "stick-pusher" system, which automatically pushes the control column or wheel forward to prevent a stall from occurring, independent of pilot action.

In order to reduce the risks in test aircraft, with marginally recoverable or non-recoverable stalls, a tail parachute is installed in some aircraft during stall development testing. When deployed, the drag from the parachute provides a nose down pitch, to allow speed to build up and aid recovery. The tail parachute must be reliably deployed, and released on demand.

Using an upward thrusting rocket as installed in the tail of the prototype de Havilland Canada Buffalo aircraft employs a different philosophy. The rocket, if fired for only a few seconds, would give sufficient thrust to lift the tail and push the nose down to assist recovery from a deep stall. Never, was it necessary to fire this rocket during stall tests on the Buffalo.

I have sat in the right hand (co-pilot's) seat on many stall tests on many aircraft and the actions sometimes displayed by, pilots while attempting to regain control and achieve a recovery from a stall, are worth describing. As the airspeed decreases towards the stall, the airflow over the wing and control surfaces become disturbed as it detaches and reattaches.

This creates a lot of buffeting and transient control forces are felt at the control wheel and rudder pedals. The pilot's hands and feet are in a whirlwind of motion, as he tries to counteract the transient forces. Recovery action is usually prompt and the workload to regain control can be extremely high. The pilot's hands and feet are very busy until control is regained. This of course, is not the type of recovery action the test pilot is hoping to eventually finalize and certify.

Take Off Speeds.

Civil transport as well as military transport aircraft must undergo a variety of tests to ensure safe take off procedures. When a passenger notices the aircraft nose come up as the pilot pulls back and rotates the control column, prior to getting airborne, little does he realize the amount of work done to obtain the speeds that are safe for take off.

On every take off, the assumption is made that an engine may fail at a critical time. This failure must not prevent the pilot from controlling the aircraft safely both on the ground and in the air. The aircraft must be able to either brake to a stop on the runway or climb away, at a required rate, with an engine failed, in the worst possible condition i.e. maximum weight, maximum airfield altitude and highest possible temperature. Mistakes made by commercial pilots, are also anticipated and tests, such as rotating the nose up too early or rotating the nose up too high or too rapidly, are undertaken to ensure safety, even under these abuse conditions. An accident occurred many years ago on a de-Havilland Comet aircraft, when the aircraft became airborne early at a high take off angle. Being in this high-drag condition, he was unable to accelerate or climb away and consequently crashed. Since that accident, all large transport aircraft are required to conduct minimum unstick speed tests to ensure this cannot occur again.

To conduct these tests the aircraft must be set to the maximum possible angle during takeoff. In order to accomplish this safely, the tail section of the fuselage is prepared for the tests by installing a friction scraper in anticipation of the tail end of the aircraft scraping the runway for a long while. Minimum unstick speed tests are conducted with the nose held high and with the tail scraping the ground. These tests are conducted at both maximum engine power and at a reduced power, representing high and hot conditions.

Tests to determine the minimum control speeds, while still on the runway, are conducted by cutting a critical engine (left outboard or right outboard, whichever gives worst results). The aircraft must not deviate more than 30 ft across the runway after the engine is cut and significantly less for narrow runways. For these tests, in order to simulate wet or icy conditions on the runway, the nose wheel steering must be disabled.

This is because nose steering would provide little contribution to control on wet or icy runways.

Tests to determine the minimum control speeds in the air after take off, are conducted by cutting the critical engine at progressively lower speeds until the minimum speed is established. Minimum control speed chosen, is a speed that will prevent the aircraft, after an engine failure, deviating more than a 20 degrees heading change with a bank angle not more than 30 degrees. In other words, the aircraft must be in reasonable control.

In addition to the above tests, stalling speeds are also required to determine take off speeds.

Resulting from all these tests, the crew will be given the following take-off speed information: V1, Vr and V2, which are presented in the aircraft flight manual.

V1 is the decision speed, which is a speed below which, if an engine fails, the aircraft must abort the take-off and brake to a stop. Above this speed, if an engine fails, the aircraft must continue the take-off and climb away.

Vr is the speed at which the pilot pulls back on the control column to initiate rotation into the air.

V2 is the target speed the pilot aims for if he continues the take-off with an engine failed.

Aircraft Stability and Control.

Aircraft stability can be defined as the ability to maintain an attitude and inherently restore that attitude after being displaced by external forces.

Aircraft controllability is the ability to direct the movement of an aircraft to change attitude and speed.

A multitude of tests at different weights and center of gravity are conducted and complex analysis is undertaken to prove that pilot's workload meets an acceptable standard.

Aircraft Performance.

The measurement of aircraft performance is one of the most important tasks to be conducted. It affects both safety and marketability. Airfield Performance Tests. Before describing performance tests, I will describe the variety of equipment used to measure ground speed and runway distance.

Photo Theodolite

A photo theodolite is a device which accurately records the angular position of a moving object. The theodolite operator tracts the object and angular position is fed into a central computer. At least two other operators, at different surveyed locations, track the same object. Photographs of the object are taken, so that any tracking errors can be corrected.

A target marked on the aircraft, is used as the tracking target during takeoff, landing ,and aborted takeoff tests. Geometric analysis provides distance, height, speed, and acceleration or deceleration data.

Photo theodolites were only available at a few military test sites and were used mainly to track bomb and rocket missile trajectories. They were expensive and did not allow any flexibility in test location. We used them for an area navigation exercise and also for water-born performance of a floatplane Twin Otter used as a water bomber.

Fairchild Performance Camera.

Developed by the Fairchild Aircraft Company, this unique camera produced equivalent data as the tracking theodolites but provided great flexibility. The camera had a fixed glass recording photo plate. A narrow vertical shutter would click on and off as the lens followed the aircraft.

Each click of the shutter, would also record an accurate time base.

When the glass plate was developed, about twenty-four images of the aircraft would be seen on the glass plate, each with a recorded time.

Analysis of the glass plate gave all the necessary data.

Nose camera

A wide-angle camera with a constant shutter speed was mounted in the nose of the aircraft and took photos of cones (the same as traffic cones) spaced two hundred feet along each side of the runway. Four cones on each photo were digitized and entered into a computer. Geometric analysis by the computer provided all necessary data.

Trisponder

The trisponder operates on the Doppler system and records distance and speed from a transmitter, to a reflector. The transmitter, with its antenna, is located in the aircraft and the reflector is located on a tripod two hundred feet off the end of the runway, on the runway centerline.

The trisponder distance and radio altimeter height above ground Is recorded on the aircraft test instrumentation. Data analysis provides all necessary information.

Gyro stabilized platform

This is a very expensive piece of equipment, used for an inertial navigation systems, consisting of sensitive and accurate accelerometers mounted on a gyro-stabilized platform. It measures, longitudinal, vertical and lateral accelerations and provides information of the aircrafts geographic position.

The data output of this navigation system, is constantly adjusted and updated, by reference to other navigational aids. Using an accurate time base, all necessary data can be obtained. I have no experience of the accuracy of this method.

Laser tracking

This is also an expensive piece of equipment, which contains all the devices required for recording and analysis. It is mounted on a vehicle.

A laser reflector is mounted on the aircraft. The laser in the vehicle automatically and accurately follows the reflector in the aircraft. Geometric analysis of recorded data is obtained in the vehicle.

Global positioning system

Once Global Positioning (GPS), using satellite technology, became available and was relatively inexpensive, it replaced all previous methods. To gain the accuracy we required, an additional satellite transmitter was placed on a tripod located on a carefully surveyed spot near the runway. Now the position of the aircraft could be recorded with an accuracy measured to an inch.

This is the method I used after my retirement when conducting contaminated runway tests in North Bay Ontario.

Deviation from the runway centerline

In order to determine the minimum control speed while on the ground, subsequent to an engine cut, deviation across the runway had to be obtained. The maximum allowable deviation was 30 ft.(Except for narrow runways).

The Global Positioning method described previously could easily conduct this task, but was not available previously, so we used the following methods, which were crude, but gave adequate results.

Slab Method.

The concrete runway we used was made with ten-foot square slabs, with their edges clearly defined. After cutting the critical engine, an observer inside the aircraft would note the sideways movement by counting the concrete slabs. A 1.75 slab count would be a 17.5 feet deviation, which was a typical number. Since deviations never went above a 2-slab (20 ft) count, this method was considered satisfactory.

Stone Method

The stone method was introduced by Transport Canada during acceptance testing. A stone-man with a pocket full of stones would be located on the runway threshold, directly behind the aircraft. He dropped a stone behind the point that the aircraft started to deviate. He then dropped a second stone directly behind the position of maximum deviation. The measurement between the stones defined the deviation.

The stone man was sometimes bowled over by the aircraft's slipstream. He soon learnt to crouch down with his back to the aircraft at the start of the take off, turning around in time to lay his first stone.

Water Trail Method

It became evident that an improved and a more accurate method was required, especially when narrow runway certification required a maximum deviation less than 6 feet.

I purchased a pressure spray unit from a garden supply store and fastened it beside the test engineer's station in the aircraft, which happened to be a de Havilland Dash 8. A rubber hose, with an on/off tap was connected to a narrow brass pipe, which trailed on the ground below the fuselage and was located between the two undercarriage legs.

The tests were conducted in Arizona, where the water soon evaporated and had to be examined and measured quickly before the evidence of the track disappeared. Several observers stationed at the runway sideline and armed with tape measures would run out, take their measurements and return to the sideline, ready for the next run. The pressurised water tank, was constantly filled and pumped up to maintain water pressure.

On a test run, the water trail could be clearly seen and both the start of the deviation and the maximum deviation could be measured from the runway centerline. This became our test method.

Video Method.

This method replaced the Water-Trail method to avoid the use of spotters on a busy runway. It provided a worms eye view of the proceedings and could rapidly provide runway deviation information.

CHAPTER 28

Airfield Performance

For performance testing, wind speeds measured on a six foot high tripod, must average not more than five miles per hour before tests can commence Airfield performance tests, are often curtailed due to wind conditions. The early morning is often the best time to conduct these tests. On many days, the test crew would keep a continuous eye on the wind. If it fell near the target wind, the crew would scramble to the aircraft, taxi to the take off point and often return as the winds increased. Airfield performance tests can be frustrating, demanding a lot of patience.

All take off tests are conducted with the critical engine cut at the critical speed. The speed at fifty feet above the runway must be not less than the safety speed (V2) recommended in the flight manual.

Distance to unstick (wheels off the ground) and distance to thirty-five feet above the ground are measured.

All aborted take off tests (Referred to as accelerate-stop tests) are conducted in the same manner as take off tests, except that, when the engine is cut, the aircraft is rapidly brought to a halt on the runway. The distance to stop is recorded.

Landing tests are conducted, both with all engines operating and with a critical engine failed prior to landing. At a height thirty feet above the runway the air speed must be thirty percent above the stall speed, as recommended in the flight manual. Below thirty feet, engine power may be reduced and after the aircraft lands it is brought to a halt as quickly as possible. Distance from thirty-feet above the runway and distance from touchdown to stop, is recorded. Distance recorded is increased by sixty percent before being entered in the flight Manual.

Tests with brake failures and with failure of the anti-skid system, are conducted to ensure that distances in the flight manual are met.

Climb tests.

Many climbs are recorded, with different flap settings and with the aircraft both at a heavy weight and at light weight.

Take off climbs are conducted, in the take off configuration, with the critical engine out and with the airspeed set to the recommended take off safety speed. Tests are conducted with undercarriage both up and down.

Other series of climbs are conducted at different flap configurations, with all engines operating, with one engine out, and two engines out, if the aircraft has more than two engines.

Test weights are tightly controlled for heavy weight tests and the aircraft is constantly returned to base for re-fueling.

Tests must be conducted in calm air conditions. Many tests are abandoned due to some turbulence encountered.

Climb tests can be quite boring when compared to take off, landing, and handling qualities tests.

As a result of all these tests, the aircraft Flight Manual is prepared.

Flight Manual

The Flight Manual, by law, must be carried aboard the aircraft and provides some of the following items:

Air speed error.

Stall speeds.

Decision speed, below which, if an engine fails, the aircraft must be brought to a stop on the runway, and above which the take off must continue.

Rotation speed. The speed at which the pilot pulls back on the elevator control to pull the nose up.

Take off safety speed. The speed targeted immediately after becoming airborne.

Approach speeds and climb speeds and climb performance..

W.A.T. Limits. (Weight, Altitude, Temperature). Data is provided to ensure adequate rate of climb after an engine failure on take off under all weight and atmospheric conditions. Also, it must be shown that an adequate rate of climb is provided, if a go around is initiated during a landing.

Take off and Landing distances are provided for all aircraft weights and for all airfield winds, temperatures and altitudes.

Maximum Operating Speed and Never Exceed Speed.

Normal and Emergency operating procedures.

Note: all the speeds presented in the flight manual provide adequate margins of safety at even the most critical of times and ensure that the aircraft speed is well above both stalling speed and minimum control speed.

High Speed Flight.

When I was a teenager, I remember seeing a movie about a test pilot. The only task he seemed to undertake was to place the aircraft in a near vertical dive. Prior to all his tests, he would always take some chewing gum out of his mouth and slap it on the rear end of his aircraft.

During the dive, there was a tremendous screaming sound as the air rushed by. The aircraft shook so much that all the cockpit instruments were a blur and the force to pull up from the dive required a Herculean effort. Just before hitting the ground our hero recovered from the dive. This indicated that the aircraft was cleared for operation.

Many different aircraft were tested in the same manner. The movie ended when the ace test pilot forgot to dab his chewing gum on the rear end of his aircraft and was unable to pull up from the dive, ending up in a ball of fire.

High speed testing now play a small but important part in aircraft tests.

Because of the potential for mainframe or control surface flutter and severe buffeting, these tests are treated with extra caution.

Before arriving at de Havilland Canada, a Caribou military transport aircraft encountered such severe flutter that the pilot's control wheel came off from his control shaft and had to be thrown aside. Both crewmembers bailed out successfully.

Often maximum speed tests are conducted with a chase plane in attendance. The chase plane could check on any control surface damage or loose and/or missing panels. I hesitate to mention it but the chase plane could also be used to spot any potential wreckage. The use of Telemetry, if available, is often used for dive tests, with a number of engineers on the ground checking on satisfactory control damping at each test speed. Test speed is then increased in small increments before advancing towards the maximum dive speed. Each control surface is subjected to a disturbing pulse. In the early days of high speed testing, the pilot sharply pulsed each control, by a short sharp tap on the elevator, aileron and rudder. The time required for damping out control oscillations, after each pulse, gave confidence in advancing to the next speed.

It is currently the practice to install a flutter vane, which will input a variety of frequencies and amplitude disturbances to the aircraft and controls.

If the aircraft can advance to the maximum dive speed without encountering control flutter and, severe buffet or vibration, the test can be considered completed. A speed, considerably lower than the dive speed demonstrated, will be entered in the flight limitation section of the flight manual, as a Never Exceed Speed. The limitations will also list a Maximum Operating Speed, which is significantly below the Never Exceed Speed. A speed warning horn is set to sound an alarm, if speed is in excess of the Maximum Operating Speed. Thus, speed limitations speeds listed in the Flight Manual have a large safety margin built in. They are well below the test dive speed that has been demonstrated.

Brake Demonstration Tests at Maximum Kinetic Energy.

Wheel brakes get heated during use and brake pads eventually wear out of tolerance. Wear indicators are provided to inform maintenance crew when brakes need changing. Brakes are designed to satisfy a maximum braking energy. Normally, maximum brake energy is not encountered, however, at the highest and hotest airfield to be operated on and with a tail wind, ground speeds during a take off run can be very high. If an engine fails at decision speed requiring the pilot to abort the takeoff and brake to a stop, the energy put into the brakes will be very high indeed and maximum braking energy will be demanded.

A test is required to demonstrate the safety of operation under maximum speed braking. Worn brakes are used for the tests. A landing is conducted and after a required delay, the aircraft must taxi a prescribed distance and line up at the start of the runway, thus the brakes will be warm before the test starts. The runway to be used must be very long to cater to the nature of the test. A very high decision speed is targeted, to ensure Maximum Kinetic Energy is achieved.

In anticipation of potential problems, a fire truck is dispatched to the end of the runway. On board the truck a test engineer will call for action, if required. With all the energy put into the brakes, the brakes are likely to catch on fire. The brakes may disintegrate and hydraulic fluid could leak on extremely hot brakes causing a significant fire. The test requirements assume problems will occur and also assume that normally, a fire truck could take 5 minutes to reach the scene. At the end of the test. If a fire occurs at the end of the test and can be ignored for five minutes, the aircraft passes the test.

It was a very difficult judgment call to be made by the test engineer on board the fire truck.

Maximum kinetic energy tests were conducted on the de Havilland Dash 8 aircraft. It was dusk when the test started and was quite dark, soon the flames coming from the brakes were spectacular; demonstrating the amount of energy used to stop. The emergency fire crew had been sitting in the fire hall for many months, maybe many years. Perhaps their only activity was putting out practice fires. They were given a briefing prior to the test and were asked to refrain themselves from going into action on any fire, except when called upon. They were also advised not to approach the fire at 90 degrees to the wheels. This was because fuse plugs, several of which were installed on the side of the wheels, could blow from overheating and thus prevent tires from bursting. If the plugs blew they would be ejected at bullet speed. After the aircraft came to a stop, the fire truck rushed to the scene. The engineer started his stopwatch as soon as the aircraft came to a stop. He was constantly gazing between his stopwatch and the localized fire on both the left and right wheel area. After five minutes, fires were still present but had started to subside. The fire crew, were then given the nod and the small fires were put out. We did pass the test.

Statistical data has shown that the chances of an engine failing at decision speed, is one in ten million. The odds of an engine failing at decision speed when operating at highest possible airfield, with the hottest possible temperature, at a maximum allowable weight and operating in a tail wind is significantly greater than one in ten million.

I hope airline passengers realize the extent of testing that is undertaken to protect them from even an extremely unlikely scenario.

Fuel System Tests

Some fuel system tests are worthy of note. Negative "g" tests require special preparation to prevent all engines shutting down due to a potential lack of fuel, as fuel moves up and away from fuel supply pipes. Only the test engine is left unmodified, the remaining engine or engines are provided with a small nitrogen-pressurized fuel tank. The nitrogen-protected tank is selected immediately prior to maneuvering to obtain the five seconds of negative required for the test. If the test engine failed to function during the five seconds of operation under negative "g", modifications to the fuel system would be required. None were required.

Later, a similar test on a Dash 7 four-engine aircraft, gave a surprising result when all four propellers went into the feather position. This episode has been previously described.

The unusable fuel in each fuel tank must be known. Due to the odd shapes of the tanks, fuel can gather in nooks, crannies, and baffles and cannot be drained to the last drop to supply the engines with fuel.

Tests are conducted one tank at a time. A small amount of fuel is placed in the test tank and the aircraft is set in a climb, using maximum continuous power, until the engine malfunctions due to fuel starvation. The test tank is isolated and the engine is restarted with fuel from another tank. On return to base, the remaining fuel in the test tank is drained into a container and weighed. Tests continued on the same tank with the aircraft in a left turn, then a right turn. Each time, the remaining fuel was weighed. The same procedure was repeated using each and every tank on the aircraft. The maximum weight recorded for each tank was identified as unusable fuel.

This surplus fuel in each tank is dead weight and must be added to the basic weight of the aircraft. This can reduce the aircraft payload, sometimes as much as the weight of a passenger. The fuel tank content indicator is required to be reset to zero with the unusable fuel in the tank. It was a lot of work just to obtain unusable fuel.

Noise tests.

Local residents living in close proximity to airports introduced external noise standards as a result of objections. Sideline and overhead noise measurements are taken during take off and landing at precise locations on the airfield. A variety of noise frequencies are compared with an acceptable noise decibel standard. Failure to meet the standard can result in expensive engine, propeller, or air intake modifications. In some cases, aircraft have operational restrictions enforced, such as a reduction in take off power.

Icing tests.

All Transport Category aircraft must demonstrate safe operation in severe icing conditions. Severe icing conditions are defined by water droplet size, liquid water content, and rate of ice build up. Special instrumentation is required to measure these parameters. Before entering icing conditions it is prudent to examine operation in icing, under controlled conditions, while checking the engine, propeller and aircraft de-icing systems. On the Buffalo aircraft, we used a United States tanker aircraft. Water replaced fuel in the tanker and a large trailing spray ring, replacing the fuel nozzle, used for air re-fueling. The ice spray could be controlled to meet the icing specification. Due to wake turbulence behind that large tanker aircraft, it was difficult for Bob Fowler, our pilot to hold steady in the somewhat narrow ice spray. We did manage to dip the left wing; engine, and propeller, in the ice spray, long enough to be satisfied that we were ready to search for natural severe icing conditions.

A Tanker aircraft was not available for the Dash 7 and Dash 8 icing program. We used the helicopter ice spray rig operated by the Canadian National Research Council. To get to the spray rig, the aircraft was towed down a steep winding incline on a rough unpaved road. It was a tricky process, indeed. When the wind was strong enough and pointing in the right direction, the aircraft was engulfed in the necessary ice spray. This gave some confidence that engines, intakes, and propeller, would not pose problems during natural icing tests.

Searching for severe natural icing conditions is never easy and often requires good luck. Meteorological forecasts are continuously studied and a dash to find ice often results in an unsuitable ice encounter. When the required quality and quantity of ice is encountered, photographic records of ice build up all around the aircraft are obtained and departure to a clear area allows the conduct of climb and handling tests with the accumulated ice.

De-icing system failures are also investigated. Ice shape build up is examined when a section of the wing, rudder or tail plane de-icing system is deliberately turned off. Photos of these shapes, are compared with calculated ice shapes. Calculated maximum encounter ice shapes, made of wood, are installed one at a time, on each flight on a critical section of the leading edge. For example, on the Dash 7, the critical ten-foot outboard wing leading edge section, approximate ten feet long had a ridiculous looking shape attached to it. I will never forget my first impression on seeing the Dash 7 ready for flight, to test the outboard de-icing boot failure. The normal smooth and rounded leading edge, was shaped like a six-inch wide molar tooth, bluntly facing into wind. I remember thinking, if we could fly with this much abuse to a leading edge of a wing, we could fly with a barn door instead of a wing. We assumed there would be a very high stall speed with this add-on and took appropriate precautions by taking off at speeds 30% above normal.

Taking off and conducting handling tests including stalls, with a wing given such abusive treatment, appeared insane, yet that is what we did.

Fear not if an aircraft encounters icing and be rest assured that the most severe icing situation has been thoroughly investigated.

Contaminated Runway Tests.

This series of tests are devised to ensure that slush or standing water on a runway does not pose a hazard during take off or landing. A water trough is built sufficiently wide to accommodate both nose wheels and main wheels. The trough must average a half-inch deep with a tolerance of plus or minus a quarter of an inch. The length of the trough can be anywhere between two hundred and fifty feet and seven hundred feet. The aircraft conducts high-speed runs into the water in the take off and landing configuration at increasing speeds, causing spectacular spray patterns. Photographic evidence is obtained of critical areas such as engine intakes, propellers, flight controls, and flap actuators.

Official Certification Tests.

When all tests are completed and reports issued, the certification authority e.g. The CAA in the UK, the FAA in the US and Transport Canada, sends an evaluation team consisting of a test pilot and several specialist engineers to examine the aircraft and test results. Resulting from this examination, a series of flight tests are undertaken and witnessed by the Authority, who re-examine critical and or marginal handling, performance, and system tests. After satisfactory completion of these tests, a representative production aircraft is required to conduct a function and reliability series of tests while conducting an agreed number of flight hours, averaging a total of two hundred hours. These flights are conducted on representative routes and include many emergency functions and system failure demonstrations. The Authorities also witness these tests.

A demonstration of an emergency evacuation on the ground by both crew and passengers is required. The evacuation drill is made as real as possible. It is conducted in the dark, using airline steward personnel and volunteer passengers complete with representative on board baggage. A small percentage of passengers are even required to carry dolls representing babies. Suitable padding is placed below the emergency exits to prevent possible injury. One of the emergency exits is rendered inoperative for added realism. Just before the evacuation begins, the hangar lights are turned off and emergency lights are turned on in the aircraft. Baggage from overhead bins is thrown into the aisle. Timing starts when the order to evacuate is given and stops when the last passenger exits the aircraft.

Every effort has been made to ensure the safety of the air traveler. Regulations to improve safety, after any rare incident, ensure even greater safety. Relax and enjoy your next flight. It's far safer than driving a car.

It is written.

Thanks be to the Gods

Acknowledgements

I wish to thank my three daughters, Linda, Susan and Anne, who, after listening to some of my stories, have been goading me to write about my varied experiences. Also to my wife Myrna, who keeps pointing out areas in the book that requires changing in order to enable the average reader to comprehend the complexity of aviation.

I also wish to thank all my colleagues, who have taught me much, and thanks also to Bob Fowler for his sage advice and Roy and Marilyn Madill who, for a while, disrupted their lives to aid me with editing. Also my thanks go to Melody Richardson, the writer in residence at the Bracebridge Public library and her mother Rose, whose encouragement and advice enabled me to complete this book.

Thanks also to the Photographic and Public Relations Departments of the Companies with whom I worked, for providing me with some of the photographs entered in this book.